Universe Dynamics

Universe Dynamics

The Least Action Principle and Lagrange's Equations

Jacques Vanier

in collaboration with

Cipriana Tomescu
Université de Montréal

CRC Press is an imprint of the
Taylor & Francis Group, an **informa** business

CRC Press
Taylor & Francis Group
6000 Broken Sound Parkway NW, Suite 300
Boca Raton, FL 33487-2742

CRC Press is an imprint of Taylor & Francis Group, an Informa business

Printed on acid-free paper

International Standard Book Number-13: 978-1-138-33589-9 (Hardback)

International Standard Book Number-13: 978-1-138-33579-0 (Paperback)

Library of Congress Cataloging-in-Publication Data

Names: Vanier, Jacques, 1934- author. | Tomescu, Cipriana, author.
Title: Universe dynamics : the least action principle and lagrange's equations /
Jacques Vanier; in collaboration with Cipriana Tomescu (Universitâe de Montrâeal).
Description: Boca Raton, FL : CRC Press, [2019] |
Includes bibliographical references and index.
Identifiers: LCCN 2018041985| ISBN 9781138335790 (pbk. : alk. paper) |
ISBN 9780429443473 (ebook : alk. paper) | ISBN 9781138335899 (hardback : alk. paper)
Subjects: LCSH: Least action. | Lagrange equations. |
Mechanics. | Physics. | Quantum theory.
Classification: LCC QA871 .V28275 2019 | DDC 531/.11–dc23
LC record available at https://lccn.loc.gov/2018041985

Visit the Taylor & Francis Web site at
www.taylorandfrancis.com

and the CRC Press Web site at
www.crcpress.com

Contents

Preface

The purpose of this short book is to offer students and knowledgeable readers an elementary introduction to the important role and use of the least action principle and the resulting Lagrange's equations in the analysis of the dynamical behavior of objects in the universe. In the majority of publications, that subject is introduced within the context of classical mechanics in several sections dispersed in a main text that is focused on the application of Newton's principles, and on occasions the laws of electromagnetism. Such an approach is generally aimed at briefly informing the reader that the description of the dynamical behaviour of the universe can also be made in a way different from the standard vector analysis based on Newton's principles and Maxwell's equations in classical mechanics. The Lagrangian approach is thus most often introduced, in various parts of the main text, to re-enforce the standard vector analysis made in most subfields. In such a presentation, it is difficult for the reader to acquire a general, coherent overview of the basic idea behind the Lagrangian approach and its impact on the understanding of the underlying physics. The present short monograph is aimed at being a coherent text, which, after reviewing in an elementary way the standard analysis used in classical physics, electromagnetism and quantum mechanics, concentrates on the physics involved in the introduction of the least action principle and Lagrange's equations and their use in several examples to describe the dynamical behavior of the universe. First-grade algebra and calculus are used throughout and the text should be accessible to most readers possessing an elementary background in the use of those mathematical tools. The text should be used as a complement to classical textbooks. It offers a short, concise outline of the Lagrangian approach in the analysis of the universe dynamics in selected subfields.

Acknowledgments

We wish to thank Pierre E Vanier, who has challenged one of the authors over the years regarding the mechanical aspect of the evolution of the universe. Some of the concepts analyzed in the present text are concerned with questions raised in the discussions that took place and we hope it provides at least a partial answer to them.

CHAPTER 1

Introduction

The study of the universe at the large scale is undoubtedly one of physics' most fascinating subjects. It concerns studying the behavior of what we see above our head, which offers us a fascinating spectacle in clear sky at night. It is most probably one of the first objects of contemplation that made human beings realize their minuscule size within that vast landscape. Throughout millennia of their existence, human beings, although becoming increasingly conscious of their nature, could not fully understand what that landscape made of moving luminous objects was. It took millennia of mental adventures, of abstract reasoning often in erroneous directions, before some appropriate rules or laws were formulated to describe approximately the behavior of material objects in space. Before those days of increased understanding, supernatural causes were assumed to be instrumental in the dynamic evolution of the universe and the motion of objects it contained. It was a long time indeed before appropriate measurement instrumentation was developed to experimentally determine the correctness of those formulated rules. In fact, it is only a few hundred years ago that geniuses such as Copernicus, Galileo and Newton, among many others, were able to make sense of the observations and measurements made relative to the behavior of those objects by formulating principles that appeared to be guiding their motion. Newton provided us with a rather subtle description of how the universe functions at the large scale. Following Galileo, he enunciated the postulate of inertia, which effectively recognized that rest and uniform motion are natural states of objects in a given frame of reference. The concept of mass was used to quantify the content of matter that an object is made of. He then introduced the concept of applied force on that mass, which was capable of changing its state of inertia. The concepts of action and reaction, the basis of some of our modern-day technology, was also introduced. He expressed those thoughts in terms of basic principles that were written as laws that can be used to describe the motion of celestial bodies and objects near us. He also introduced the law of gravitational attraction, which describes how material objects behave in the presence of one another. Those laws are expressed mathematically, that is in a subtle language invented by human beings. That language allowed us to verify both qualitatively and quantitatively to great accuracy the validity of the laws in relation to our observations. This approach is very different to the philosophical approach, which is based on abstract reasoning, leading to so-called theses that cannot be verified by means of measurements and numbers. With the advent of the theory developed by Newton, we could make measurements and, in particular, make predictions relative to the observation of most phenomena, such as the motion of objects near the surface of the earth and planetary orbits. As intelligent beings, we are very proud of that achievement.

We should still try to remain modest, however. Although we can describe well physical phenomena and their dynamic evolution, we essentially do not know *why* nature works the way it does. Are those mathematical laws real in the sense that nature obeys them absolutely in its evolution? Or are they simply an invention of our mind as we attempt to understand that evolution and make approximate predictions on the future dynamical behavior of objects? Is the universe using our language or a language of its own that we do not yet have access to? We do not know. The laws, as expressed in our mathematical language, result from observations and intuition. We do not know much about the *inner working gears* of that universe. For example, distant material objects appear to attract each other, but we don't know why they do so without even touching each other. What is the "*ethereal fluid*" that flows from one object to the other to signal its nearby presence and create that so-called gravitational force of attraction? Newton raised the question but could not provide an answer. Within that context, the best we can do is to describe the phenomenon using our mathematics and invent concepts such as gravitational potential and gravitational field that create the force of attraction. Einstein addressed the question in a brilliant way. In his general theory of relativity he provided a step towards a better understanding of how gravity works. His theory describes the large scale mechanical behavior of the universe extremely well. It is essentially based on the existence of a new entity called *spacetime*, which was created by combining space and time into a single four-dimensional frame of reference, with mass causing a curvature of that spacetime. In that theory, spacetime is characterized by a universal constant c, which is the speed at which electromagnetic waves travel in all inertial frames of references, whatever their relative speed. The theory proved the validity of Newton's principles as an approximation for essentially all applications in our daily activities. However, it resulted in a total change of our perception of space, time and gravitation itself.

We can make similar remarks about several other subfields of physics and their evolution. We do not really know *why* the universe works the way it does, but we observe phenomena and develop mathematical theories to explain *how* it works the way it does. For example, the phenomena originating from the presence of electrical charges is rather mysterious. Nevertheless, over the years the development of our understanding of such phenomena has been outstanding. Gauss, Faraday and several others have proposed various independent laws that describe the behavior of those charges in the presence of one another. These descriptions culminated in the magnificent theory developed by Maxwell, known as *electromagnetism*, a somewhat frightening appellation, which simply groups into a single framework our knowledge on phenomena observed when electrically charged objects are in the presence of each other or move relative to each other. Again, the concepts of potential and field were introduced and seem to describe all observations extremely well. The theory developed is outstanding in providing a mathematical description of how observed electrical and magnetic phenomena behave. Another example is the theory of quantum physics, which is probably the most mysterious of all, and which is based essentially on an observation that defies common sense: material particles appear to sometimes behave as particles, and other times as waves. Matter has a dual personality. That observation has led to a rather complex description of the physics of particles at the small-scale level. It takes concepts developed in the theories mentioned above – mechanics, electromagnetism, relativity – and couples them to some so-called quantum rules or principles, showing that these theories are interlaced and really basic attempts to mathematically describe the functioning of the whole universe. In more recent years, the theory has developed into the so-called quantum field theory in which all particles are considered as originating from the excitation of so-called quantum fields, similar to the case of electromagnetic waves, whose quantization led to the creation of the photon concept. To complete this enumeration of the main subfields of physics, we should also mention the theories that describe the inner working of the atomic nucleus at the ultra-small scale. They address the interaction between constituting nuclear particles, a subfield generally called particle physics. This requires the introduction of new forces, called weak and strong, and describes, in a rather complex way, however, observations that are made at that ultra-small scale.

Many treatises have been written on each of those subjects. It is difficult to add to these treatises except perhaps by presenting the subjects in a different way. In a previous book by one of the authors, called *The Universe, a Challenge to the Mind* (Imperial College Press 2011), a description of the functioning of the universe as a whole was given in simple language, without using mathematics. That book was an attempt to introduce the reader to the laws that appear to govern the dynamical behavior of the universe at both the small and large scales. In general, however, since the field of physics as a whole is rather vast and complex, subfields are studied in separate specialized treatises. We say that the field is compartmented. This approach does not cause problems, because theories associated with specific subfields are specialized and it is possible to treat a given subfield with a minimum of overlap with other subfields. Of course, it has been a challenge for the scientific community to find a unified fundamental theory that would provide a single basis for all subfields such as classical mechanics, electromagnetism, quantum theory, particle physics and relativity. This would be a theory of everything, sometimes abbreviated to TOE. Unfortunately, the challenge is great, and so thus far the goal has not been reached. Is such a theory possible in the mathematical context we have developed? We do not know. The question has always been of great interest since the early development of physics. It is still today the subject of great efforts in the theoretical physicists' community. However, in the search for such a theory, common basic principles underlying the functioning of the universe were discovered. We may think, for example, of the principle of conservation of some specific entities when changes are made in the physical environment where observations are made. A particular one is the principle of conservation of energy. It is the first law of thermodynamics and is not disputed. It is one of the most fundamental laws and one that is used often as an element of the development of many theories within subfields. However, it appears that it results from a more fundamental principle related to symmetry and from deeper considerations on the fundamental nature of the universe. Is there a common principle that forms the basis of those conservation principles? Is there, in our mathematical language, a law that we can accept as a basic principle that appears to dictate how the universe behaves dynamically?

In the present text, we will try to highlight the fundamental way nature seems to work, obeying an intuitive principle that, when examined more carefully, touches upon, although loosely, the fundamental question of why the universe behaves the way it does. We may introduce it immediately. The universe in its dynamical behavior appears to look for a stationary state in which symmetry plays a major role. *It is the principle of least action.* This behavior appears to be present in all subfields that have been the subject of detailed theories. This finding does not lead to a unified theory of the whole of physics but at least it provides a basis for unifying our thinking in working in physics subfields. It further opens doors in the imagination of humans to venture into finding new approaches that could explain the behavior of the universe in particular situations.

We know that there are four fundamental types of interaction that generate forces and dictate the dynamics of the universe. These are: the force of gravity, electromagnetic forces, weak interaction and strong interaction. In the present monography, we will first conduct an elementary review of the essential elements of the subject known as classical physics as covered in the first years of university physics courses. We will also review some basic concepts that lead to quantum physics. Then, we will introduce two concepts: *action* and *Lagrangian.* Within that context, we will introduce the *least action principle,* which leads to the so-called differential *Lagrange's equations.* Those equations prescribe the form that the Lagrangian must have in order to properly describe the dynamics of the universe. They are sometimes assimilated to equations of motion. The approach does not lead to new physics. It does not teach us *why* the universe works according to that simple principle, but it gets us closer to an understanding of the dynamical behavior of the universe we live in.

The least action principle has effectively been applied to and is used in essentially all subfields of physics. This is generally done in the following way: a Lagrangian is first assumed to represent the particular situation studied and then Lagrange's equations are applied to it. The result is then analyzed within the context of known physical laws and observations. In the present text, however, we will limit ourselves to a few subfields, such as classical mechanics, electromagnetism, relativity and elementary quantum physics. We will not elaborate on the field of particle physics. That field, including nuclear forces, is extremely complex and falls outside the context and goal of the present short elementary text, which is to introduce the newcomer to the field of Lagrange's approach in the simplest way possible.

In the present text, mathematics will be the language used, but we will try to make the physics explicit and accessible. Nevertheless, the text requires that the reader has basic knowledge of differential calculus, vector algebra and matrix algebra, as well as an openness towards some acrobatic demonstration. The elementary approach used will restrict us to the more elementary material and, sometimes, we will have to simply accept some results. However, we will always attempt to highlight as much as possible the physics involved.

On the other hand, we have attempted to reduce the complexity of the mathematics that is involved in the derivation of the various subjects. We have made use of many textbooks and several general articles available in open literature. In many cases it is not possible to simplify the mathematics involved. Sometimes, the results of advanced calculations must be accepted as such. There are only a few ways that basic principles can be introduced. This will certainly be perceived in many sections of this text. Nevertheless, the beginner may find in this text a coherent presentation that may be useful as an introduction to the more advanced books already published and listed at the end of the text. He may also find this text to be a good introduction to more advanced fields such as, for example, quantum field theory, which has led to the standard model of particle physics.

CHAPTER II

Selected Elements of Classical and Quantum Physics

In the present chapter we will review, in an elementary way, basic theories that describe the dynamical behavior of objects and particles with mass and electric charges. We will outline the general results obtained with the standard approaches used in classical mechanics, relativity, electromagnetism and quantum physics. The review will not amount to a rigorous treatment of these areas. The reader seeking more detailed treatment of those fields is referred to classical texts enumerated at the end of the present book.

In the 18th century, Newton revolutionized our approach to attempting to understand the mechanical functioning of the universe. Among many other things, he developed a theory that *explains* the motion of material objects without electric charges. Later, theories were developed to explain or describe the behavior of charged particles, leading to Maxwell's equation, which fully describes phenomena such as electricity and magnetism and the propagation of electromagnetic waves. Still later, a theory was developed by Schrödinger to explain the behavior of particles at the atomic level. The exercise led to the creation of quantum theory.

The word *explain* is emphasized here in italics to draw attention to the limit of the theories that we introduce. Theories generally do not say *why* material bodies behave in space the way they do. In fact, theories just explain *how* things behave. In this sense they provide a description of their behavior by means of mathematical analysis. The existentialist question of why the universe behaves the way it does, or even exists, has not been answered yet and probably will never be. We are an integral part of the landscape and due to our myopic view of that universe, we cannot have a so-called *God's view* of the whole. It looks very much like we may never know what that whole thing is about. We are very limited. In practice, we simply observe a given phenomenon and we develop a so- called logical "theory" to describe what we observe. To do so, we use the language of mathematics, but we do not know if the universe uses that same language to evolve. And we do not know why it works the way it does. We have developed that mathematical language to describe the phenomena we observe, and we use it to make predictions relative to other connected phenomena. If the results agree with the predictions, we claim that our theory is logically correct. This is valid until someone else observes a phenomenon whose details are not quite explained by that first theory. It is then claimed that the theory is not exact and may be an approximation of a more fundamental one. A good example of this is the theory of relativity developed by Einstein, Lorentz, Poincaré, Minkowski and others in the 20th century, a theory that supplemented Newton's approach. It explains the results of observations obtained by means of accurate physical measurements, for example, the alteration of space and time as observed for an object in motion, the deflection of light rays by large masses, the precession of elongated planetary orbits, the existence of gravitational waves and observations of other phenomena that could not be explained by means of the earlier theory developed by Newton in the 18th century.

In this chapter, we will first outline a general framework that will provide us with some basic concepts and mathematical tools used in the description of those theories. We will then outline the basic theory that was introduced by Newton relative to the behavior of objects in motion in our universe and we will review essential concepts introduced in relativity, electromagnetism and quantum physics.

A. FRAMEWORK

We can base our understanding of the behavior of objects in our universe, observed at the large scale, on four entities that intuitively play a major role: they are *space, time, mass* and *electric charge*. We will first elaborate on various concepts and mathematical tools that form a framework for the analysis of the dynamics of those objects or particles.

Space, Systems of Coordinates

We live in a surrounding that we call space and objects move in that surrounding. Space appears to us as having three dimensions and objects evolve in it. We represent a portion of that space in Figure 2.1 by means of a so called three-dimensional (3-D) Cartesian system of coordinates.

This system is composed of three axes, x, y and z, which are orthogonal to each other. The whole system thus consists essentially of three planes intersecting each other at right angles. The crossing point of these planes is identified as the zero of the three axes. An object of mass m is introduced in the figure at position x_1, y_1 and z_1. The object may be assumed to move at velocity **v**. In the same figure, we also show another system of coordinates that we will use: polar coordinates r, θ and φ. The object is situated at the end of vector $\boldsymbol{r}$. Discrete values of x, y and z, or r, θ and φ, are sufficient to situate an object of mass m in that space.

In the Cartesian system, the coordinates are further qualified by means of unit vectors **i, j** and **k** parallel to the x, y and z axes. We

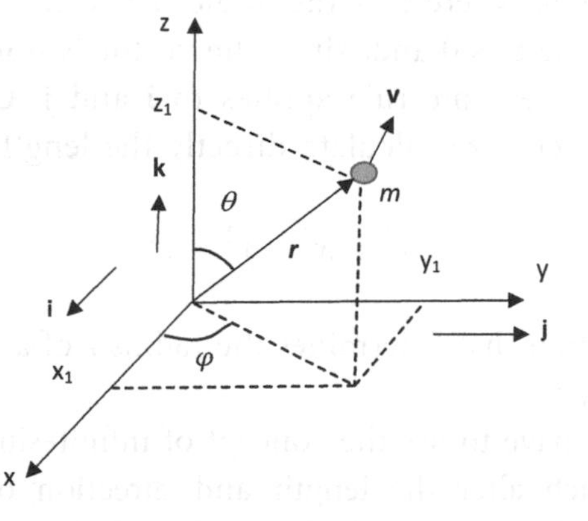

FIGURE 2.1 3-D system of axis used in visualizing the basic principles described in the text. We have included in that figure Cartesian and polar coordinates. Their relation is outlined in the main text.

may also introduce basic unit vectors for polar coordinates $\mathbf{e_r}$, $\mathbf{e_\theta}$ and $\mathbf{e_\varphi}$. The relation between the Cartesian and polar coordinates is obtained directly from the figure. We have in general:

$$\boldsymbol{r} = \mathbf{i}x + \mathbf{j}y + \mathbf{k}z \tag{2.1}$$

and

$$x = r sin\theta\ cos\varphi \tag{2.2}$$

$$y = r\ sin\theta\ sin\varphi \tag{2.3}$$

$$z = r cos\theta \tag{2.4}$$

The vectors **i**, **j** and **k** are orthogonal. Using vector algebra, we have the obvious relations:

$$\mathbf{i} \cdot \mathbf{i} = 1,\ \ \mathbf{j} \cdot \mathbf{j} = 1,\ \ \mathbf{k} \cdot \mathbf{k} = 1, \tag{2.5}$$

$$\mathbf{j} \cdot \mathbf{i} = \mathbf{i} \cdot \mathbf{j} = \mathbf{k} \cdot \mathbf{i} = \mathbf{i} \cdot \mathbf{k} = \mathbf{j} \cdot \mathbf{k} = \mathbf{k} \cdot \mathbf{j} = 0 \tag{2.6}$$

where bold characters are used to identify the vector property of a quantity and a *bold dot* means a scalar product. For example, **k·k** means (k k) *cos* θ, where θ is the angle between the two vectors. In that case the angle θ is 0 and, since the vector is normalized to unity, the result is 1. The same rule applies to **i** and **j**. Using Eq. 2.1 and Pythagoras theorem, we calculate directly the length of vector $\boldsymbol{r}$ as:

$$r^2 = x^2 + y^2 + z^2 \tag{2.7}$$

This is the equation that determines the radius r of a sphere in terms of coordinates x, y, z.

We will often have to use the concept of infinitesimal increments dx, dy and dz, which alter the length and direction of vector $\boldsymbol{r}$ by an increment $\boldsymbol{dr}$. It is important to know the relations that exist between these various increments because we will use them below. To more easily understand the process of calculation we will calculate such relations in a 2-D system first and then extend it to a 3-D coordinates system. Such a 2-D system is shown in Figure 2.2. It is essentially a flat plane, like the surface of a sheet of paper.

We assume the presence of a mass m at position (x_1, y_1) in that plane. Its position is given by the vector:

$$r = \mathbf{i}x_1 + \mathbf{j}y_1 \tag{2.8}$$

Its length, as was calculated above for the 3-D system, is given by:

$$r^2 = x_1^2 + y_1^2 \quad \text{or} \quad |r| = \sqrt{x_1^2 + y_1^2} \tag{2.9}$$

We introduce the concept of the *line element* $d\mathbf{S}$ describing the change in position of the mass or its displacement:

$$d\mathbf{S} = \mathbf{i}dx + \mathbf{j}dy \tag{2.10}$$

The square of the length of this displacement is given by the scalar product:

$$dS^2 = d\mathbf{S} \cdot d\mathbf{S} = \mathbf{i}dx \cdot \mathbf{i}dx + \mathbf{j}dy \cdot \mathbf{j}dy + \mathbf{i}dx \cdot \mathbf{j}dy + \mathbf{j}dy \cdot \mathbf{i}dx \tag{2.11}$$

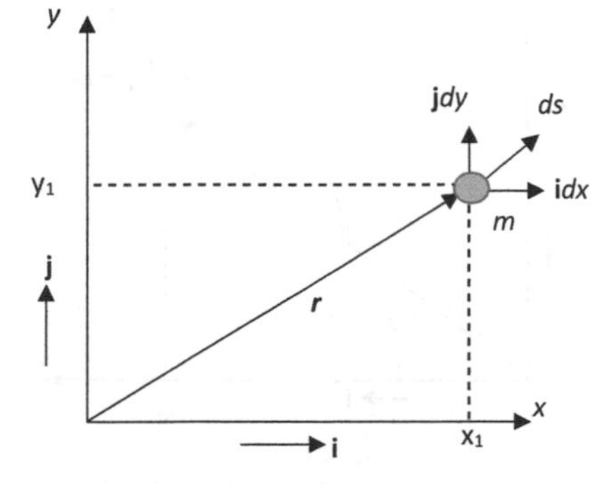

FIGURE 2.2 2-D Cartesian system used to define the increments dx and dy.

which according to Eq. 2.6 gives:

$$dS^2 = dx^2 + dy^2 \quad \text{or} \quad dS = \sqrt{dx^2 + dy^2} \tag{2.12}$$

To situate our mass in the 2-D flat plane, we could have used a polar coordinate system as shown in Figure 2.3.

We can interpret the previous analysis in terms of those new coordinates r and θ. We call the basis vectors along r and θ in that new system $\mathbf{e}_\theta$ and $\mathbf{e}_r$. The line element is then written as:

$$dS = \mathbf{e}_r dr + \mathbf{e}_\theta r d\theta \tag{2.13}$$

and we have

$$(dS)^2 = dS \cdot dS = dr^2 + r^2 d\theta^2 \tag{2.14}$$

We can conduct the same type of analysis for the case of a 3-D coordinate system. The line element in Cartesian and polar coordinates is:

$$dS^2 = dx^2 + dy^2 + dz^2 \tag{2.15}$$

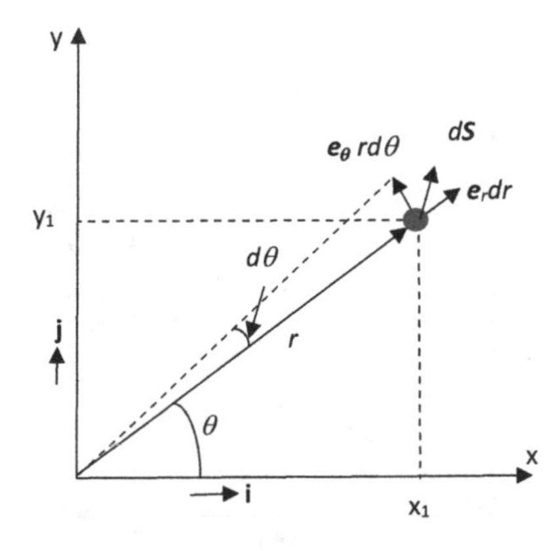

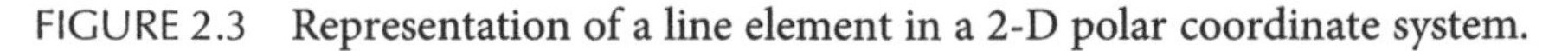
FIGURE 2.3 Representation of a line element in a 2-D polar coordinate system.

$$dS^2 = dr^2 + r^2 d\theta^2 + r^2 \sin^2\theta \, d\varphi^2 \tag{2.16}$$

These expressions are often used in analysis involving changes such as rotations of coordinates systems in 3-D space.

Rotation and Translation of Axes

Let us assume two 3-D systems of axes Σ and Σ' rotated relative to each other as shown in Figure 2.4.

In system Σ, the length of vector r is given by:

$$x^2 + y^2 + z^2 = r^2 \quad \text{or} \quad \sum x_i^2 = r^2 \; (i = 1,\ 2,\ 3) \tag{2.17}$$

in which we have introduced the notation x_1, x_2, x_3 often used for x, y, z.

Similarly, in Σ', vector $\boldsymbol{r}$ is given by:

$$x'^2 + y'^2 + z'^2 = r^2 \quad \text{or} \quad \text{still} \sum x_i'^2 = r^2 \tag{2.18}$$

The length of the vector is the same in both systems and we can write

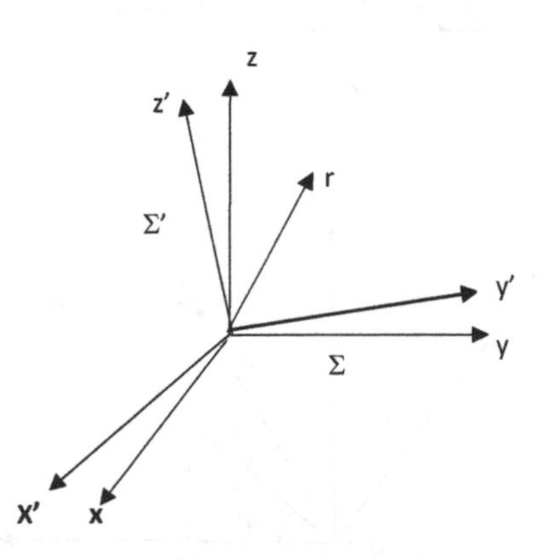

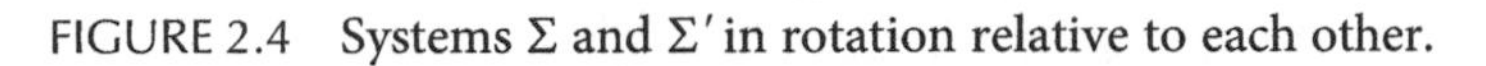

FIGURE 2.4 Systems Σ and Σ' in rotation relative to each other.

$$\sum x_i'^2 = \sum x_i^2 \quad (\mathrm{i} = 1,\ 2,\ 3) \tag{2.19}$$

We can foresee a unique relation between coordinates in Σ and those in Σ′. Such an exercise leads to a relation in matrix form:

$$[x_i'] = A[x_i] \tag{2.20}$$

where A is the matrix of the transformation of the axis. This equation can be written explicitly as:

$$\begin{bmatrix} x_1' \\ x_2' \\ x_3' \end{bmatrix} = \begin{bmatrix} a_{11} & a_{12} & a_{13} \\ a_{21} & a_{22} & a_{23} \\ a_{31} & a_{32} & a_{33} \end{bmatrix} \begin{bmatrix} x_1 \\ x_2 \\ x_3 \end{bmatrix} \tag{2.21}$$

Coefficients a_{ij} are projection coefficients of an axis in one system on a corresponding axis in the other system. They are a trigonometric function connected to the axis of rotation. They are the functions to be determined.

The calculation is somewhat lengthy, and to illustrate the technique we will do a detailed calculation of the case of two 2-D systems rotated relative to each other by angle φ, as shown in Figure 2.5.

We thus have:

$$\sum x_i'^2 = \sum x_i^2 \quad (i = 1,\ 2) \tag{2.22}$$

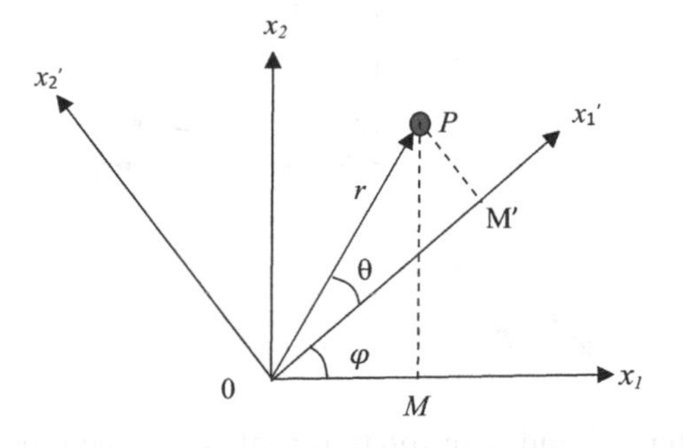

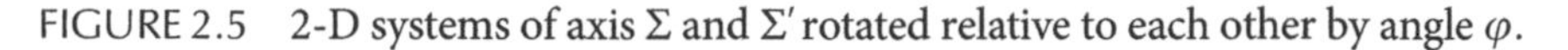
FIGURE 2.5 2-D systems of axis Σ and Σ′ rotated relative to each other by angle φ.

with

$$x_i' = \sum_j a_{ij} x_j \quad (i = 1,\ 2) \tag{2.23}$$

We replace Eq. 2.23 into Eq. 2.22 and the left-hand member becomes:

$$\sum_i x_i'^2 = \sum_i \sum_j a_{ij} x_j \sum_k a_{ik} x_k \tag{2.24}$$

We rearrange the terms and we obtain:

$$\sum_i x_i'^2 = \sum_i \Big(\sum_{jk} a_{ij} a_{ik} x_j x_k \Big) \tag{2.25}$$

This must be equal to the right-hand side of Eq. 2.23 with j=k. Thus we have:

$$\text{for } j = k \qquad \sum_i a_{ij} a_{ik} = 1 \tag{2.26}$$

$$\text{for } j \neq k \qquad \sum_i a_{ij} a_{ik} = 0 \tag{2.27}$$

In a 2-D system, the transformation matrix is written as:

$$A = \begin{bmatrix} a_{11} & a_{12} \\ a_{21} & a_{22} \end{bmatrix} \tag{2.28}$$

We can evaluate coefficients a_{ij} with the help of Figure 2.5. The length of the vector is the same in both systems. It is OP. The projection components on the axis of both systems are:

in Σ':

$$OM' = x_1' = OPcos\theta \tag{2.29}$$

$$M'P = x_2' = OPsin\theta \tag{2.30}$$

and in Σ

$$OM = x_1 = OPcos(\theta + \varphi) \tag{2.31}$$

$$MP = x_2 = OPsin(\theta + \varphi) \tag{2.32}$$

The following trigonometric relations hold:

$$\cos(\theta + \varphi) = cos\theta cos\varphi - \sin\theta sin\varphi \tag{2.33}$$

$$\sin(\theta + \varphi) = sin\theta cos\varphi + \cos\theta sin\varphi \tag{2.34}$$

We replace those in Eqs. 2.31 and 2.32 and take into account Eqs. 2.29 and 2.30. We obtain:

$$x_1 = x_1' cos\varphi - x_2' sin\varphi \tag{2.35}$$

$$x_2 = x_1' sin\varphi + x_2' cos\varphi \tag{2.36}$$

We use Eq. 2.20 in our 2-D system,

$$[x_i] = A[x_i'] \tag{2.37}$$

and the coefficients are:

$$A = \begin{bmatrix} cos\varphi & -sin\varphi \\ sin\varphi & cos\varphi \end{bmatrix} \tag{2.38}$$

Similarly, we have:

$$[x_i'] = A^{-1}\, x_i \tag{2.39}$$

where A^{-1} is the inverse matrix obtained by exchanging rows and columns in A. Thus:

$$A^{-1} = \begin{bmatrix} cos\varphi & sin\varphi \\ -sin\varphi & cos\varphi \end{bmatrix} \tag{2.40}$$

and we have:

$$AA^{-1} = \begin{bmatrix} 1 & 0 \\ 0 & 1 \end{bmatrix} = [1] \tag{2.41}$$

where [1] is the unit matrix.

The question of translation of one system relative to the other is rather simple when no rotation is involved. If system Σ' is in motion at uniform velocity $\mathbf{v}$ relative to Σ, we can re-orient the axes of one of the two systems and line up the motion direction to a particular axis, say axis z or z′. Such an approach simplifies calculations of relative motion and the relation between the axes becomes:

$$z' = z + vt, \qquad x' = x, \qquad y' = y \tag{2.42}$$

This amounts to what is called a Galilean transformation.

Elements of Vector Algebra

In this text, we will use vector algebra very often. We have just used some of the properties of vectors like scalar product and projection of a vector on an axis. In this section we will recall some other properties and associated theorems. We recall that a vector is an entity that has magnitude and direction. Speed, force and momentum, which we will introduce shortly, are vectors. On the other hand, a scalar is a physical entity that has magnitude only, for example mass. We will also introduce the concept of field and potential. A field may be a scalar with magnitude but without direction. Temperature, electrical and gravitational potentials are scalar fields. A field may also be a vector with magnitude and direction such as an electric or a gravitational field.

In Cartesian space, we have, as mentioned above in the case of unit vectors, products of vectors F_1 and F_2 such as:

$$\boldsymbol{F_1} \cdot \boldsymbol{F_2} = \textit{scalar product} = \textit{dot product} = |F_1||F_2|\cos\theta = \textit{scalar} \quad (2.43)$$

where θ is the angle between the two vectors. We have used this definition earlier with unit vectors at a right angle to each other.

Another product of two vectors, called a vector product, is defined as:

$$\boldsymbol{F_1} \times \boldsymbol{F_2} = \textit{vector product} = |F_1||F_2|\sin\theta = \textit{vector} \quad (2.44)$$

where θ is the angle between the two vectors.

A so-called vector operator, $\boldsymbol{\nabla}$, called "del" is also introduced. In Cartesian coordinates, it is defined as:

$$\boldsymbol{\nabla} = \mathbf{i}\frac{d}{dx} + \mathbf{j}\frac{d}{dy} + \mathbf{k}\frac{d}{dz} \quad (2.45)$$

When applied to a scalar function f or a vector function $\boldsymbol{F}$, we define the following operations and results:

$\boldsymbol{\nabla} f$ A direct operation. The result is a vector called a *gradient.*

$\boldsymbol{\nabla} \cdot \boldsymbol{F}$ A scalar operation. The result is called *divergence* and is a scalar.

$\boldsymbol{\nabla} \times \boldsymbol{F}$ A vector operation. The result is called *curl* and is a vector.

$\nabla^2 = \boldsymbol{\nabla} \cdot \boldsymbol{\nabla}$ This is called the *Laplacian.*

The following theorems also apply:

1. Gradient theorem: assume that f represents a potential that varies over space as in Figure 2.6A. In the figure the intensity of the potential is represented by the density of the background, a more intense potential in the darker region.

The difference of potential between the two points 1 and 2 is then given generally by:

$$f_2 - f_1 = \int_1^2 \boldsymbol{\nabla} f \cdot dl \quad (2.46)$$

where $\boldsymbol{\nabla} f$ is the gradient of the potential f along path l.

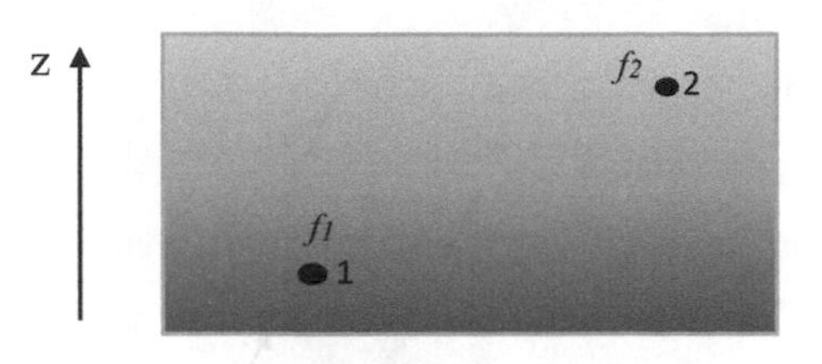

FIGURE 2.6A Potential f represented by means of the density of the background varying linearly with axis z.

2. Divergence theorem: assume that $\boldsymbol{F}$ is a vector field within a volume V of surface S as shown in Figure 2.6B.

The following relation holds:

$$\int_V \nabla \cdot \boldsymbol{F}\, d\mathrm{v} = \oint_S \boldsymbol{F} \cdot \mathrm{d}\boldsymbol{s} \tag{2.47}$$

where $\nabla \cdot \mathbf{F}$ is the divergence of $\boldsymbol{F}$. This is called the divergence theorem. This may be interpreted as follows. For example, the volume V may contain a source of radiation that has to exit through surface S: this is measured by the integral on the left. The surface integral on the right measures what comes out of the volume. It must be equal to the radiation emitted by the internal source. If that is zero, there are no sources of radiation in volume V.

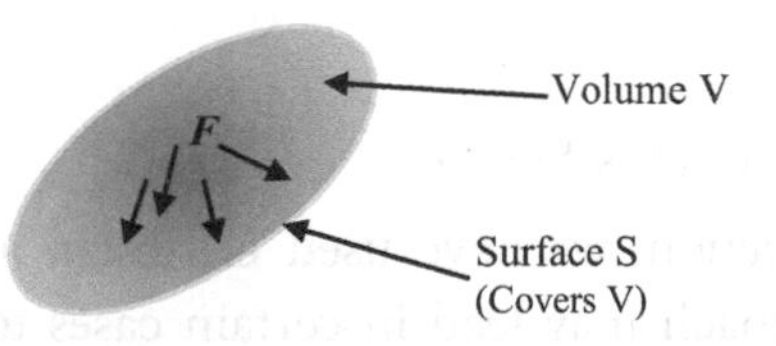

FIGURE 2.6B Representation of a field $\boldsymbol{F}$ within a volume V limited by surface S.

3. Stoke's theorem. Assume $\boldsymbol{F}$ is a vector field on surface S limited by a curve C as shown in Figure 2.6C.

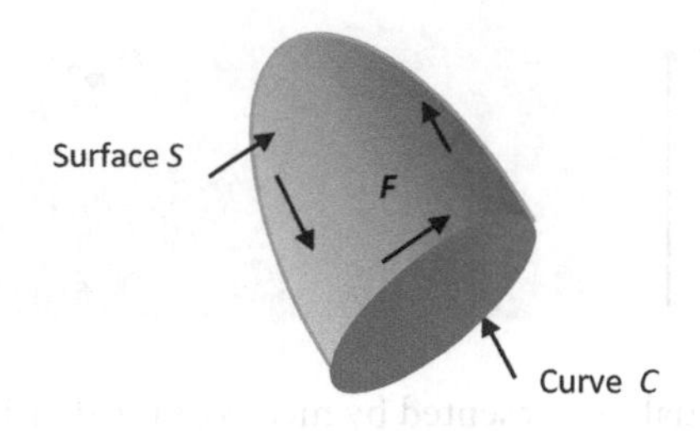

FIGURE 2.6C Representation of a vector field $\boldsymbol{F}$ on surface S limited by curve C.

The following relation holds:

$$\int_S \nabla \times \boldsymbol{F}\, ds = \oint_c \boldsymbol{F} \cdot \boldsymbol{dl} \tag{2.48}$$

where $\nabla \times \boldsymbol{F}$ is the curl of $\boldsymbol{F}$. It is called Stoke's theorem. It simply connects properties of $\boldsymbol{F}$ on the surface to its properties at the limit of that surface.

4. Finally, if $\nabla \times \boldsymbol{F} = \boldsymbol{0}$, $\boldsymbol{F}$ may be written as the gradient of a potential ϕ, that is $\boldsymbol{F} = -\nabla\phi$. This property is used to introduce the concept of *potential*.

Demonstration of those theorems can be found in many textbooks, especially those covering the field of electromagnetism (see for example G. P. Harnwell, *Principles of Electricity and Electromagnetism*).

Generalized Coordinates System

In the preceding section, we have used Cartesian and polar coordinates. However, that approach may lead in certain cases to situations where the analysis becomes extremely tedious. To avoid that problem, we will often use the concept of *generalized coordinates*. This concept is usually introduced in the context of a system composed of a large number of particles or objects, with mass m, whose positions are characterized by their individual Cartesian coordinates, x, y and z. If there are N particles, we are in the presence of $3N$ coordinates. If the coordinates are not connected to each other, they can take any value independent of other coordinates. We say that we are in the presence of *3N degrees of freedom*. This may make the study of the system evolution almost impossible when N is large. However, if there are *constraints* on the motion of the particles and if some of the coordinates or their relation to each other are fixed, like in a solid object, relative motion is impaired. The number of *independent coordinates* necessary to study the system evolution may then be considerably reduced. We then say that the number of degrees of freedom is reduced. These independent coordinates define a so-called *configuration space* in which the system evolves. To simplify the analysis of the system evolution, it is best to use a minimum number of coordinates. Consequently, it is best to introduce parameters that form a system of coordinates that take advantage of those constraints and define the configuration space appropriately. We call those *generalized coordinates. The* letter q_i has been adopted at large in the literature as the symbol representing such coordinates.

On the other hand, in a general configuration space, those coordinates need not be spatial coordinates. For example, they could be speed, momentum, potential or still energy, or a mixture of any of these. In any case, the main idea to remember when using those coordinates is to choose those that simplify most the calculation. Consequently, one needs to select generalized coordinates in a configuration space that characterizes the system in a complete a way as possible and simplify calculations as much as possible. Thus, instead of representing the position of a particle by x, y and z in Cartesian space, we may represent it in configuration space by generalized coordinates q_i. These represent the number of degrees of freedom of a given particle identified as α. The position vector r_α of that particle, which may be a function of time t, may be written in terms of generalized coordinates as:

$$\boldsymbol{r}_\alpha = \boldsymbol{r}_\alpha(q_1, q_2, q_3 \ldots, t) \tag{2.49}$$

Since $\boldsymbol{r}_\alpha$ can also be written in terms of x, y and z in Cartesian space, each of the Cartesian coordinates is then a function of the q_i, and there are transformation rules between them, rules that depend on the generalized coordinates selected.

Using the general rules of differential calculus, we can derive several relations connecting r_α to the generalized coordinates, relations that we will use extensively later. For example, a small change δq_i of generalized coordinate q_i will create a small change, δr_α, of r_α by the quantity:

$$\delta r_\alpha = \sum_i \frac{\partial r_\alpha}{\partial q_i} \delta q_i \tag{2.50}$$

in which the derivative $\frac{\partial r_\alpha}{\partial q_i}$ is just the rate of change of r_α caused by a change in q_i. The summation is over all coordinates q_i. In the equation shown, it is assumed that time is not involved, and the displacement is said to be *virtual*. A real displacement involves a time change. We will develop similar relations when required.

Time

We see objects moving in space. We qualify that motion by means of the parameter "*time*" t. We are so familiar with it that we do not think much about its properties. But what is time? It is a concept that we, human beings, have invented to be able to describe changes that we observe in our universe, such as motion of material objects. It is a concept that we cannot really define and so we simply observe its existence. We age and we observe that time flows. If time did not exist, it is difficult to imagine the universe and our existence in it. We can situate ourselves in space by specifying the value of x, y and z coordinates and we can move in that space, thus altering x, y and z. However, we are at the mercy of time. *It is increasing all the time.* In relativity theory, we will introduce time as an axis. We simply take the product of the speed of light c by time t, call it axis ct, which has dimension of space, and we combine space and time to form a single entity called *spacetime*. However, although the axis has the dimension of space it does not behave at all like space. In space we can move in all directions back and forth, but in time we only go forward independently of what we do. For the moment, we will limit ourselves to a Cartesian three-dimensional space, represented by a so-called 3-D coordinate system with axes x, y, z, as shown in Figure 2.1. We will then use time t as a continuous parameter to describe motion of objects in that space. Sometimes, we will use the notation x_i where i varies from 1 to 3 for x, y, z, or from 1 to 4 representing x, y, z and t.

Velocity and Acceleration

If the particle is moving at constant velocity **v**, its motion is determined totally by the derivative of vector $\boldsymbol{r}$ with respect to time:

$$\mathbf{v} = \frac{d\boldsymbol{r}}{dt} = \dot{\boldsymbol{r}} \tag{2.51}$$

We will use the notation of a *dot* on a parameter to represent its derivative in respect to time. The position of mass m shown in Figure 2.1 is totally specified by the three Cartesian coordinates. During the motion of m in space, the value $\boldsymbol{r}$ changes and all x, y and z coordinates may change. We say that the particle has three degrees of freedom.

On the other hand, if the velocity of the particle is changing with time due to an external cause, we say that the particle is accelerated. This is usually represented by the vector $\boldsymbol{a}$ given by the derivative of **v**:

$$\boldsymbol{a} = \frac{d\mathbf{v}}{dt} = \ddot{\boldsymbol{r}} \tag{2.52}$$

Velocity can also be expressed in terms of generalized coordinates. If $\boldsymbol{r}_\alpha$ is itself a function of time, the velocity $\mathbf{v}_\alpha$ of particle α may be written as:

$$\mathbf{v}_\alpha = \dot{\boldsymbol{r}}_\alpha = \frac{d\boldsymbol{r}_\alpha}{dt} = \sum_i \frac{\partial \boldsymbol{r}_\alpha}{\partial q_i}\frac{dq_i}{dt} + \frac{\partial \boldsymbol{r}_\alpha}{\partial t} \tag{2.53}$$

or

$$\mathbf{v}_\alpha = \sum_i \frac{\partial \boldsymbol{r}_\alpha}{\partial q_i}\dot{q}_i + \frac{\partial \boldsymbol{r}_\alpha}{\partial t} \tag{2.54}$$

in which the derivative of $\boldsymbol{r}_\alpha$ relative to q_i is interpreted as above as the rate of change of $\boldsymbol{r}_\alpha$ with q_i.

Mass

The concept, called mass, is introduced to quantify the amount of matter in an object of a given volume. However, we find that it is much more than that and that it raises some fundamental questions relative to its exact nature. In fact, we can interpret mass in various ways, for example we can say that mass is an entity:

- at the basis of Newton's second law,
- that creates inertia,
- that creates attraction of material massive objects between themselves,
- that causes spacetime curvature,
- that behaves like energy,
- ...

One main question regards the rules that govern the evolution of masses in time and space. These rules were formulated by Newton in the 17th century. They form the center of what we call in physics the theory of *classical mechanics*. That theory is outstanding. It explains essentially all the observations we can make in our daily activities. It has provided the basic tools to send a man to the moon and bring him back.

The theory can be presented in a simple form using vector algebra. However, as we will see in the following chapters, it can also be developed in a more advanced mathematical context, (*Lagrange's equations*) using an intuitive principle (*least action principle*). Using that approach, at first sight the mathematics will appear a bit abstract and frightening, but it is very rational and the analysis leads to rather simple conclusions in agreement with observations. We will outline that more advanced approach in the following chapters using a style that, we hope, will be accessible to most readers with a basic background in mathematics and physics. For the moment, we will first develop the standard approach using Newton's elementary principles and vector algebra. We do not plan to cover the whole field of classical physics. We will limit our analysis to only a few selected topics in order to introduce the reader to the field and we will use mathematics in as simple a way as possible.

B. UNIVERSE DYNAMICS USING NEWTON'S AND GALILEO'S PRINCIPLES

In the following paragraphs, we will introduce in a very elementary way the basic principles that are necessary to describe the dynamics of macroscopic objects in space.

Inertia

A free material object of mass m, not affected by other particles, or in other words by external forces, moving at velocity **v**, keeps moving in space at that same velocity and in the same direction. It does not need to be pushed by some divine spirit. It simply goes on its way. In free space without obstruction or applied force, the velocity of a material object is constant in direction and value. This concept is called inertia. As illustrated in Figure 2.1, we can situate an object of mass m at position $\boldsymbol{r}$ or (x_1, y_1, z_1) at a time t_1 in a given frame of reference. If the object moves at velocity $\mathbf{v} = d\boldsymbol{r}/dt$, and no force is applied, it will continue to do so indefinitely at the same speed and same direction. The frame of coordinates shown is called a frame of reference and may also be called a frame of inertia. We note immediately that we can attach a system of reference to the mass itself and make that system move at velocity **v** with the object. In that case, if we set the new origin on the mass itself, the vector $\boldsymbol{r}$ has zero value and does not change with time. Then, $d\boldsymbol{r}/dt = \mathbf{v} = 0$ and an observer also attached to the same frame and moving with it sees that object at rest and cannot tell if he or the object is in motion or not. Consequently, we need to be careful about how we define the frame of reference in which we make our observations or to which we, as an observer, are attached.

Force and Acceleration

Let us represent the path of the particle of mass m moving at velocity **v** as shown in Figure 2.7 in a 2-D space.

It is assumed that a force $\boldsymbol{F}$ is applied and alters the velocity **v** of the particle, that is to say, its speed and direction of motion. The force may

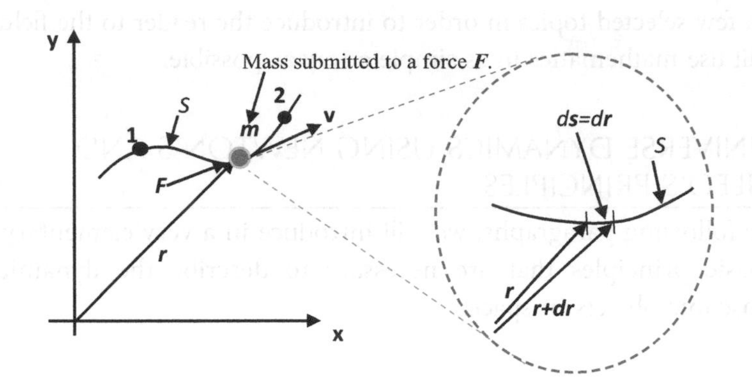

FIGURE 2.7 Representation of the path followed by a particle of mass m moving in space along a path S and submitted to a force $\boldsymbol{F}$.

change in direction and the path of the object follows a curve identified as S in the figure. The velocity is obviously tangent to the path S described by the length and direction of the vector $\boldsymbol{r}$ and we can thus represent it also as:

$$\mathbf{v} = \frac{d\boldsymbol{S}}{dt} = \frac{d\boldsymbol{r}}{dt} \tag{2.55}$$

where $d\boldsymbol{S}$ is an infinitesimal distance along path S and is equivalent to the infinitesimal change $d\boldsymbol{r}$ of the vector $\boldsymbol{r}$ during time dt.

The motion of the particle is assumed to obey the law:

$$\boldsymbol{F} = \frac{d}{dt}\,\boldsymbol{p} \tag{2.56}$$

where $\boldsymbol{p}$ may be interpreted as an entity characterizing the "*quantity of movement*" and called the *linear momentum*, equal to the product of mass m *by* $\mathbf{v}$. Thus $\boldsymbol{p} = m\mathbf{v}$. At this stage, the law given by Eq. 2.56 is not demonstrated and follows simply from intuition, logic and agreement with observation or measurements in the coordinate system chosen. It may be accepted as a basic postulate. If it is assumed that m is constant, the law can also be written as:

$$\boldsymbol{F} = m\frac{d}{dt}\mathbf{v} = m\frac{d^2}{dt^2}\boldsymbol{r} = m\ddot{r} = m\,\boldsymbol{a} \tag{2.57}$$

where $\boldsymbol{a}$ is called the acceleration of the object, as defined earlier. This is given in *meters per second squared.* According to the first principle of inertia, we note that to change the speed and direction of motion, a force is required. We may ask ourselves, what exactly is a force? In fact, Eq. 2.56 can be used either to define force or mass. When we push on an object at rest with our hands we may say that we apply a force to the object: it goes in the direction we push and the object acquires speed. What are then the phenomena at the surfaces of our hands and of the object that cause the interaction and result in a force? In fact, the origin of the interaction is electromagnetic and quantum mechanical and it is rather mysterious. Below, we will introduce the force of gravity that pulls objects with mass towards each other. We will see that it is also extremely mysterious.

If no force is applied, the velocity stays constant. Thus, the resistance to changes in velocity may also be called inertia. In fact, the law just given shows how an object of mass m opposes resistance to a change in its motion. It requires a force to alter $\mathbf{v}$.

Action and Reaction

A third rule in that theory addresses the question of action and reaction. It simply says that if a force acts on an object, there is an effect produced by that force called action and there will be a reaction to that force. That reaction acts as a force and opposes the action of the force. In equilibrium, the two forces, that applied and that opposing it, are equal and sum up to zero. For example, if one pushes on the wall of a room with his hand, the wall may move a little, but reacts by means of an opposing force. The two forces oppose each other, an equilibrium is created and the wall does not fall. This phenomenon is responsible for the recoil on the shoulder when a bullet is shot from a rifle. It is the phenomenon that makes possible the functioning of a rocket or a jet airplane. Ejection of gases by the engines creates a reaction force forward, causing acceleration of the rocket or airplane. We may even rewrite Eq. 2.57 as:

$$\boldsymbol{F} - m\frac{d}{dt}\mathbf{v} = 0 \tag{2.58}$$

This expression is rather interesting. It looks like an equilibrium equation, the reaction force $-m\frac{d}{dt}\mathbf{v}$ being opposed to and always in equilibrium with the applied force $\boldsymbol{F}$. We will come back to that point and show that the idea is an alternative way of arriving at the so-called Lagrange's equations that makes possible the analysis of motion of objects in a manner totally different from the elementary one using Newton's principle and vector algebra as is the case here. These are the basic principles that describe in simple terms the behavior of massive objects at rest or in motion in space. There are, however, several other concepts that we need to introduce in order to be comfortable in the use of those principles.

Energy

When we apply a force and accelerate an object from a speed $\mathbf{v}_1$ to a speed $\mathbf{v}_2$, we say that we do *work* on the object because of its resistance to a change in its dynamic state, which we called inertia. In the process, the work done is transferred to the object, which we then call *energy* acquired by the object. We call this energy *kinetic energy*, and we can calculate it in the following way. Assume, as in Figure 2.7, an object of mass m in motion on a path that we call S. During time dt, the object moves a distance $d\boldsymbol{S}$. The component of $d\boldsymbol{S}$, which lies in the direction of the force, is a measure of the effect of the force itself on the object motion. We define work as the product of the projection of $\boldsymbol{F}$ along $d\boldsymbol{S}$, that is $F \cos \theta \, dS$, where θ is the angle between $\boldsymbol{F}$ and the actual displacement $d\boldsymbol{S}$. This displacement is written as a scalar product, as introduced earlier, and the work element is given by:

$$dW = \boldsymbol{F} \cdot d\boldsymbol{S} \tag{2.59}$$

It is a very reasonable definition that quantifies the effort we make to displace the particle against inertia in the direction of the force. It is energy spent by us and given to the particle. We can calculate easily the work done by force $\boldsymbol{F}$ in moving the particle from point 1 to point 2 over path S of Figure 2.7. The velocity varies in direction and value. We integrate Eq. 2.59 over the path $\boldsymbol{S}$ from point 1 to point 2:

$$W = \int_{1}^{2} \boldsymbol{F} \cdot d\boldsymbol{S} \tag{2.60}$$

We use Eq. 2.57 for $\boldsymbol{F}$ and the definition of acceleration $\boldsymbol{a}$ as a derivative of $\mathbf{v}$ with respect to time. We also use Eq. 2.55 for $d\boldsymbol{S}$, giving $d\boldsymbol{S} = \mathbf{v}dt$. We integrate Eq. 2.60 from t_1 to t_2 and calculate the total work W done by us. We obtain:

$$W_{12} = \int_{t_1}^{t_2} m\frac{d\mathbf{v}}{dt} \cdot \mathbf{v}dt = \int_{\mathrm{v}_1}^{\mathrm{v}_2} m\mathbf{v} \cdot d\mathbf{v} = \frac{1}{2}m\left(\mathrm{v}_2^2 - \mathrm{v}_1^2\right) \tag{2.61}$$

This is energy acquired by the object from the agent producing the force, when its actual velocity changes from $\mathbf{v_1}$ at time t_1 to $\mathbf{v_2}$ at time t_2. Thus, the work done on the object has gone into increasing its energy from $\frac{1}{2}m\mathrm{v}_1^2$ to $\frac{1}{2}m\mathrm{v}_2^2$ by the action of the applied force. The object is now moving in the rest frame of reference at velocity $\mathbf{v_2}$, its earlier velocity being $\mathbf{v_1}$. The general term $\frac{1}{2}m\mathrm{v}^2$ is called *kinetic energy*. It is simply identified as T, or KE when notation becomes ambiguous:

$$T = KE = \frac{1}{2}m\mathrm{v}^2 \tag{2.62}$$

We note that we have used Newton's second law, $\boldsymbol{F} = m\boldsymbol{a}$, in deriving this equation and in future conclusions we should remember that.

We note also that the derivation was done using vector calculus. Thus, the expression above for T can be transformed readily into various coordinate systems such as Cartesian, polar or generalized coordinates. For a given object or particle, designated by α, we can write:

$$T_\alpha = \frac{1}{2} m v_\alpha^2 = \frac{1}{2} m \mathbf{v}_\alpha \cdot \mathbf{v}_\alpha = \frac{1}{2} m \dot{\boldsymbol{r}}_\alpha \cdot \dot{\boldsymbol{r}}_\alpha \tag{2.63}$$

where we have written $(v_\alpha)^2$ as a scalar product of $\mathbf{v}_\alpha$ by itself. Using the result obtained above, and assuming that $\boldsymbol{r}_\alpha$ does not depend on time explicitly, as is often the case, we may write this in generalized coordinates as:

$$T = \frac{1}{2} m \sum_{i,j} \frac{\partial \boldsymbol{r}_\alpha}{\partial q_i} \cdot \frac{\partial \boldsymbol{r}_\alpha}{\partial q_j} \dot{q}_i \dot{q}_j \tag{2.64}$$

We may also write in terms of generalized coordinates the virtual work done by an applied force $\boldsymbol{F}_\alpha$ causing a displacement $\delta \boldsymbol{r}_\alpha$ as:

$$\boldsymbol{F}_\alpha \cdot \delta \boldsymbol{r}_\alpha = \sum_i \boldsymbol{F}_\alpha \cdot \frac{\partial \boldsymbol{r}_\alpha}{\partial q_i} \delta q_i \tag{2.65}$$

We can then define the component $\boldsymbol{Q}_i$ of the generalized force as:

$$\boldsymbol{Q}_i = \boldsymbol{F}_\alpha \cdot \frac{\partial \boldsymbol{r}_\alpha}{\partial q_i} \tag{2.66}$$

These expressions will be used later on.

Kinetic energy is one form of energy that adds to the properties of an object in motion. However, we can introduce immediately another form of energy that will play an important role in what follows. It is called *potential energy*. To do that, we apply Eq. 2.60 to a particular situation. Suppose an object of which we are studying the motion travels on a closed path under the influence of an applied force. In such a situation, the particle returns to point 1, which now coincides with point 2 in Figure 2.7. The path of the particle is closed. If the kinetic energy of the object is the same at the beginning and the end of the path, then $T_1 = T_2$ and no work is done during the path. We thus have:

$$W = \oint \boldsymbol{F} \cdot d\boldsymbol{S} = 0 \tag{2.67}$$

This could happen, for example, in the case of a central force that remains at a right angle to the path of the object. As we will see, that would be the special case of a small object travelling under the force of gravity in a circular orbit around a large mass. Then, $\boldsymbol{F}$ and $d\boldsymbol{S}$ are at a right angle all the time and obviously the scalar product in Eq. 2.60 is equal to zero: no work is done by the force applied. For such a situation to take place there should not be any dissipative force present such as friction, for example. If friction were present, energy would be lost, being transformed into heat, and speed would be altered. If Eq. 2.67 is valid, we say that the force is *conservative*. If there were friction forces, energy would be dissipated, and the integral would not be zero: we would say that those forces are *non-conservative*. At this point we can use some more advanced vector algebra and the theorems we have introduced earlier. Actually, Stoke's theorem connecting surface and line integrals can be used to transform Eq. 2.67 to another simple vector relation. For Eq. 2.67 to be valid implies that the curl of the force $\boldsymbol{F}$ must be 0:

$$\boldsymbol{\nabla} \times \boldsymbol{F} = 0 \tag{2.68}$$

By means of vector algebra it is then shown that this new equation is valid only if $\boldsymbol{F}$ is the gradient of a scalar function V. We write this as:

$$\boldsymbol{F} = -\boldsymbol{\nabla} V \tag{2.69}$$

We call V the potential energy of the object and its variation in space creates a force $\boldsymbol{F}$. This way, another form of energy that describes the properties of an object is introduced. It can also be seen that the definition of V by means of Eq. 2.69 allows the addition to V of a term constant over space without altering the result. We recall that the potential is introduced under the condition that Eq. 2.68 is valid, resulting from the presence of conservative forces only, dissipative forces being absent.

At this stage, the introduction of V appears somewhat artificial, being introduced by vector algebra. However, it plays a very important role in essentially all of physics, in allowing us to derive basic principles that govern the dynamic evolution of the universe. We will re-introduce the concept later in another way and use it extensively.

Gravity

It is observed that objects fall to the ground, as if they were attracted by an invisible force. This is called *gravitational* force. Although we feel it rather evidently when we fall and hurt ourselves, it is the weakest force that we know of in the universe. On planet earth, we feel it as a strong force because the earth has a large mass. Newton could not identify the physical origin of such a force. He assumed that it acted instantaneously at a distance between all objects in the universe whatever their size. He assumed that it was proportional to the material content of the objects that attracted each other. We call this matter gravitational mass m_g. He also assumed that the force of attraction decreased with the square of the distance r between the objects. We thus have:

$$\boldsymbol{F}_g = -G\frac{m_{g1}\, m_{g2}}{r^2}\mathbf{e}_r \tag{2.70}$$

where $\mathbf{e}_r$ is the unit vector giving the direction of the force, which is oriented along the line joining the two objects. In the reference system used, as shown in Figure 2.8, the force is assumed negative opposite to the unit vector $\mathbf{e}_r$. Here, G is the so-called gravitational constant and has the value 6.67408x10^{-11} $m^3kg^{-1}s^{-2}$. For a small object of mass m_{g1} falling to the surface of the earth, m_{g2} becomes the mass of the earth M_e, much larger than the object mass, m_{g1}. We now know that the force does not act instantaneously and that its effect propagates in space at the speed of light. This property is at the basis of the theory of relativity. In the case of spherical homogeneous objects, the force acts as if the whole mass of each object was concentrated at their own geometrical center. The force is directed along $\boldsymbol{r}$, a vector from the center of one mass to the center of the other mass.

We must realize that it took millennia for human beings to arrive at that rule without essentially knowing why it works that way. It is not

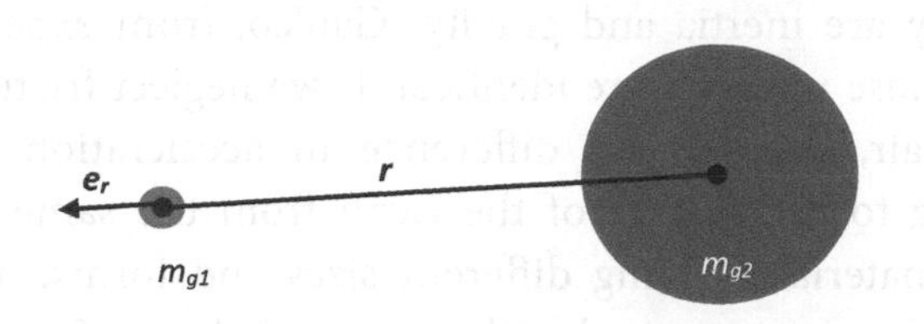

FIGURE 2.8 Two masses m_{g1} an m_{g2} attract each other with a gravitational force given by Eq. 2.70, oriented along the direction $\boldsymbol{r}$ and as if the mass was concentrated at the center of each object.

evident at all. Why is it the inverse of the square of the distance between the masses? Why is it not another power? We find that it is in agreement with observations and measurements. We do not know the gears of the mechanism that creates that law. "*Nature*" has decided it is that way. Einstein has provided an interesting explanation, establishing the effect of gravity as originating from a *curvature* of the so-called spacetime grid caused by the presence of mass and forcing objects with mass to experience and follow that curvature in their motion.

Inertial and Gravitational Mass

In the present model, an object is accelerated by the force of gravity given above. On the other hand, we have the law connected to inertia:

$$\boldsymbol{F} = m_{in}\,\boldsymbol{a} \tag{2.71}$$

where m_{in} is the inertial mass. The force of gravity is given by Eq. 2.70. We thus conclude that an object of mass m_{gr} falling towards the surface of the earth of mass M_e suffers an acceleration a in its fall to the earth's surface. We thus have:

$$m_{in}a = m_{gr}G\frac{M_e}{r^2} \qquad or \qquad a = \frac{m_{gr}}{m_{in}}G\frac{M_e}{r^2} \tag{2.72}$$

The first obvious question, of course, relates to the difference between inertial and gravitational masses, both being the content of matter in the objects in question. In fact, the physical concepts involved are totally different: they are inertia and gravity. Galileo, from experiments, concluded that those masses were identical. If we neglect friction due to the presence of air, there is no difference in acceleration between two objects falling to the surface of the earth from the same height, made of different materials, having different sizes and forms, or still having different weights, as measured with a spring balance. For example, take two balls of the same size and same weight and let them fall from a certain height. They arrive at the surface of the earth at the same time. Now, divide one of the balls into two smaller balls of half the original size, attach these three objects together with loose treads and let them fall from the same height. There will be no tension on the treads during travel and they will all arrive at the surface of the earth at the same time. It is said that Galileo had experimented this concordance from the Leaning Tower of Pisa. It was concluded that the acceleration due to gravity was identical for all falling objects and that inertial mass and gravitational mass were identical. This statement may be written as:

$$m_{gr} = m_{in} \tag{2.73}$$

although it should be recalled that the two masses represent two different entities and that the identity could be considered as resulting from a property characterizing space. This is a statement with extremely important consequences. It is part of the principle of equivalence. It says that the motion of objects in a gravitational environment is independent of their mass. There must thus be something that happens to spacetime to create that peculiar behavior: Einstein called it curvature.

In space, near the surface of the earth, at distance R_e from the earth center and neglecting the presence of other objects, the acceleration of any particle or object is given a special symbol g:

$$a = G\frac{M_e}{R_e^2} = g \tag{2.74}$$

The parameter g is defined as the acceleration of gravity at the surface of the earth and is approximately equal to 9.81 m/s^2. Thus, an object falling towards the surface of the earth, neglecting friction from the surrounding air, is accelerated at that rate independent of its nature and, in particular, independent of its mass. This is a great discovery: gravitational force is essentially characterized by a single parameter, acceleration.

In view of the large value of M_e, g can be used approximatively over a certain range of height near the surface of the earth. We can show that in the following way. Assume that an object is at height h above the surface of the earth. From Eq. 2.74, we can write g at height h as:

$$g_h = \frac{GM_e}{(R_e + h)^2} = \frac{GM_e}{R_e^2} \frac{1}{(1 + h/R_e)^2} = G\frac{M_e}{R_e^2}\left(1 - 2\left(\frac{h}{R_e}\right) + \ldots\right) \quad (2.75)$$

With $h = 1000$ m and $R_e = 6371$ km we have $\left(\frac{h}{R_e}\right) \approx 10^{-4}$. The correction for such a height is thus small and, in cases not requiring great accuracy, we can use g as a constant near the surface of the earth.

Gravitational Field

But what really is gravity? It is natural to think that *gravity* and the resulting *acceleration* have something rather particular in common. When mass is present, large or small, it appears that something happens to space and causes a change in the motion of objects close by. It seems that mass changes the nature of space. As a first step in that direction of thinking, the effect of gravity on space may be thought of as introducing a so-called *gravitational field.* We may think of the effect of a mass M as altering the surrounding space by introducing a vector field, $\boldsymbol{E}_{gr}$, given by:

$$\boldsymbol{E}_{gr} = -\frac{GM}{r^2}\mathbf{e}_r \quad (2.76)$$

For simplicity, we assume that the mass of objects is distributed uniformly over their respective spherical volumes. With the field defined as in Eq. 2.76, we can write immediately the gravitational force $\boldsymbol{F}_{gr}$ on mass m as:

$$\boldsymbol{F}_{gr} = m\,\boldsymbol{E}_{gr} = -\,m\,\frac{GM}{r^2}\mathbf{e}_r \quad (2.77)$$

giving back Newton's gravitational force. We recall that $\mathbf{e}_r$ is a unit vector along the line joining the centers of the objects and gives the direction of the field. In that case, mass may be thought of as being concentrated at the center of the objects. In the following, we will again assume that mass M is much larger than mass m and that the gravitational field is not affected by the presence of the smaller mass. In reality, when the two masses are similar in size, gravity affects reciprocally each mass and relative motion is more complicated since the motion of each object is affected by the mass of the other object. In that case, it is necessary to introduce the concept of center of mass. In the present case, we consider the small object of mass m as a probe having essentially no effect on mass M. The situation is illustrated in Figure 2.9 for the case of a perfectly spherical mass M distributed uniformly in a volume of radius R with the center at coordinate $\boldsymbol{r} = 0$.

At this point it is very useful to introduce the so-called scalar gravitational potential ϕ_{gr}, defined as:

$$\phi_{gr} = -\frac{GM_e}{r} \tag{2.78}$$

from which the field can be obtained as a variation in space or gradient of that potential:

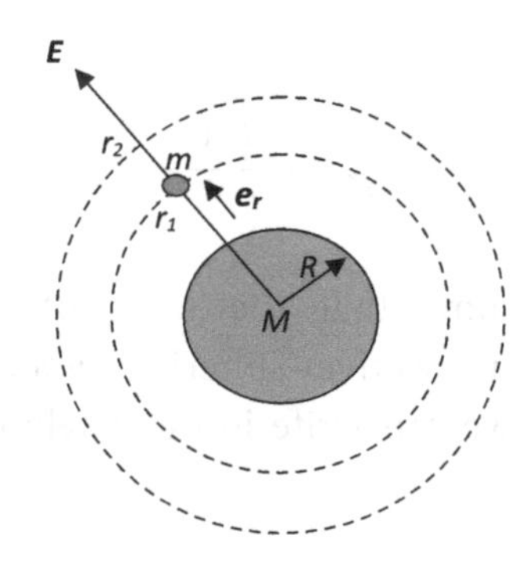

FIGURE 2.9 The gravitational field $\boldsymbol{E}_{gr}$, due to a homogeneous spherical mass is along $\mathbf{e}_r$, but is negative. It decreases as $1/r^2$

$$\boldsymbol{E}_{gr} = -\boldsymbol{\nabla}\phi_{gr} = -\frac{d}{d\boldsymbol{r}}\phi_{gr} = -\frac{GM_e}{r^2}\mathbf{e}_r \tag{2.79}$$

Comparing Eq. 2.79 to Eq. 2.74, we observe that g can be assimilated to that field. We note that since the spherical area in the space surrounding spherical mass M_e is equal to $4\pi\ r^2$, it is then natural to think of the field, its intensity being spread over the surface, as a quantity decreasing as $1/r^2$ and providing a natural basis for the force to decrease as $1/r^2$. It is most interesting to note that if for some reason the potential ϕ_{gr} does not vary with position in space, the associated field, being its derivative relative to $\boldsymbol{r}$, is then zero and the potential does not create any force on an object with mass. Consequently, a constant term may be added to ϕ_{gr} without affecting the gravitational force.

Gravitational Potential Energy

We can now introduce the important concept called gravitational potential energy V in a similar way as we have introduced it earlier in the case of conservative forces. From the definition of $\boldsymbol{F}$ in terms of $\boldsymbol{E}$ in Eq. 2.77 and the definition of work in Eq. 2.59, we may calculate the amount of work to be done for displacing mass m from position r_1 to r_2 in Figure 2.9. It is given by:

$$W = \int_{r_1}^{r_2} F \cdot d\boldsymbol{r} = \int_{r_1}^{r_2} m\boldsymbol{E} \cdot d\boldsymbol{r} \tag{2.80}$$

Using Eq. 2.79 for $\boldsymbol{E}$ we have:

$$W = m\int_{r_1}^{r_2}\left(-\frac{d}{d\boldsymbol{r}}\ \phi_{gr}\right)\cdot d\boldsymbol{r} = -m\int_{r_1}^{r_2} d\phi_{gr} \tag{2.81}$$

$$W = -m\left(\phi_{gr_2} - \phi_{gr_1}\right) = V_1 - V_2 \tag{2.82}$$

We call the term $V = m\phi_{gr}$ the potential energy of small mass m at distance r from the center of the large mass M_e. Thus, we have in general:

$$V = -m\frac{GM_e}{r} \tag{2.83}$$

The potential energy is negative, zero at large distances when r tends to infinity and is infinite when r tends to zero. For the moment, we will avoid that last situation causing a profound difficulty since, when an entity tends to infinity, our mathematics become useless.

Thus, when we apply a force to an object and move it against the force of gravity, we do work and we spend energy in doing so. The work done ΔW in elevating a mass m from distance r_1 to distance r_2 is then given by Eq. 2.82, re-written here as:

$$W = -G\,\frac{mM_e}{r_2} + G\,\frac{mM_e}{r_1} \tag{2.84}$$

In doing so we have not added nor subtracted any kinetic energy to the object of mass m. We have simply re-situated the object in two different places where it has a different potential energy. We spent energy in doing that. It all went into the increase in potential energy of the object in the new potential. We recall that gravitational potential energy is negative. Consequently, for the object at r_2, the potential energy is less negative and energy is acquired by the object. Thus, energy must be spent in elevating the object from r_1 to r_2 and it comes from the agent that has elevated the object to r_2.

As mentioned above, in view of the definition of the field $\boldsymbol{E}$ in terms of a gradient of the potential ϕ, we can add to ϕ_{gr} a constant potential ϕ_0 that does not vary in space. This is due to the fact that the gradient of a potential constant in space is 0. That constant potential does not create a force. Consequently, we can start measuring potential energy from any arbitrary position or value of r without altering the expression for the force of gravity as expressed through the field. We can set the value of the potential energy V to be zero at the surface of the earth. Then, using the approximation introduced by means of Eq. 2.75, the potential energy at small height h near the earth's surface can be written approximately as:

$$V = mgh \tag{2.85}$$

This is obtained by displacing the zero of the potential by a constant value, that is the value at the surface of the earth. It is then just a linear function, the potential energy increasing linearly with height h. Using this expression, we see that energy must be spent in elevating an object in the gravitational field of the earth in agreement with the previous derivation. However, this expression is approximate and is only valid near the surface of the earth.

An Observation

The gravitational field is an abstract concept and it raises questions relative to its exact nature. What does it do to space? As we mentioned, Newton did not attempt to explain what was transferred from one mass to the other through space to cause interaction between them without even touching each other. We have just seen that we could think of it as the effect of a scalar potential whose variation in space produces a gravitational field and the resulting gravitational attractive force. Applying Newton's law, we will find that the motion of an object orbiting another massive one is independent of its mass. This is a rather puzzling consequence of the basic principles we have introduced. As mentioned earlier, the question was addressed by Einstein in his theory of general relativity, in which he assumed that mass causes a curvature of spacetime. Objects then follow the curvature of the spacetime grid.

Gravity: Poisson's Equation

The gravitational field concept can be introduced in a slightly different mathematical approach to the one used above. Assume for simplicity that a mass M is distributed uniformly within a spherical volume V. We can introduce the concept of a so-called flux density $\boldsymbol{\Phi}$ at distance r from the mass center, defined as:

$$\boldsymbol{\Phi} = -\frac{\rho V}{4\pi r^2}\mathbf{e}_r \tag{2.86}$$

where ρ is the density of mass within volume V centered on the origin. The coordinate r is simply the distance from the origin to the point where the flux density is measured and $\mathbf{e}_r$ is a unit vector along r. The total gravitational flux over the sphere of radius r is then given by the integral over the surface of the sphere:

$$\Phi_{total} = \oint_{surf} \boldsymbol{\Phi} \cdot \boldsymbol{ds} \tag{2.87}$$

where $\boldsymbol{ds}$ is an element of surface. Within the definition made above for $\boldsymbol{\Phi}$, this reduces to:

$$\Phi_{total} = -M = -\int_{vol} \rho dv \tag{2.88}$$

We have defined the field by means of Eq. 2.76, which in the present context may be written as:

$$\boldsymbol{E} = -G \int_{vol} \frac{M}{r^2}\,\mathbf{e}_r \tag{2.89}$$

From the definition of the flux density we can thus write $\boldsymbol{E}$ as:

$$\boldsymbol{E} = G4\pi\boldsymbol{\Phi} \tag{2.90}$$

On the other hand we may use the divergence theorem, which, in the present case, can be written:

$$\oint_{surf} \boldsymbol{\Phi} \cdot \boldsymbol{ds} = \int_{vol} \boldsymbol{\nabla} \cdot \Phi dv \tag{2.91}$$

Simple algebra, using Eq. 2.87, then gives:

$$\nabla \cdot \boldsymbol{\Phi} = -\rho \tag{2.92}$$

Using Eq. 2.90, we can write finally:

$$\nabla \cdot \boldsymbol{E} = G4\pi \nabla \cdot \boldsymbol{\Phi} \tag{2.93}$$

$$\nabla \cdot \boldsymbol{E} = -G4\pi\rho \tag{2.94}$$

We note that, according to Eqs. 2.74 and 2.76 we can write this as:

$$\nabla \cdot \boldsymbol{g} = -G4\pi\rho \tag{2.95}$$

Gravity is a conservative force and we can write the field E as the gradient of the potential ϕ:

$$\boldsymbol{E} = -\nabla\phi \tag{2.96}$$

Replacing in Eq. 2.94 we finally obtain:

$$\nabla^2\phi = G4\pi\rho \tag{2.97}$$

which is the so-called Poisson's equation for the gravitational field. This result is extremely useful. It essentially expresses Newton's principles in a concise way. It also provides a starting point for deriving Hubble's constant for the expansion of the universe (see for example: Peebles, *Principles of Cosmology*, p. 421; Vanier, *The Universe*, p. 460) without using Einstein's field equations. We will use it later in our approach using the least action principle and assuming that gravity acts as a field.

Conservation Principles

We have introduced some basic concepts, laws and principles related to the motion of objects, force, work, kinetic and potential energy and we have outlined their relations. We will now use those to introduce some other concepts that are characteristic of our universe. Those concepts seem to be fundamental and even more basic than the laws that we have introduced. They are concerned with symmetry and the conservation of properties. For the moment, we will limit ourselves to some conservation laws that regulate the evolution of the universe and the motion of objects in it. We will introduce those laws in an elementary way. However, we will revisit the question in Chapter IV by means of Lagrange's equations and show, as demonstrated by Noether in the beginning of the 19th century, that these principles find their origin in symmetry properties that characterize the physics involved.

Conservation of Energy.
Energy in the universe is conserved. Energy is not created from nothing and does not disappear. There is thus an equilibrium relative to energy at large. The statement may be accepted as a postulate. Energy, as we have said may take various forms. As specific examples, we may make the following list:

- An object with mass in motion has energy.
- An object in a gravitational field has energy.
- A compressed spring has energy.
- Static fields, such as electrostatic and magnetostatic fields, are also forms of energy.
- Mass appears to be energy.
- Radiation and heat are forms of energy.
- ...

We could go on like this. What is important to realize is that one form of energy can be transformed into another form and in the process the sum of all energies involved is conserved. Nothing is lost or created in the process. It is almost evident and is a very useful principle.

Can we imagine situations where energy is not conserved? Well, yes, but this only means that we have not included in our analysis all forms of energy that are present in the phenomenon we are analyzing. For example, the presence of air in experiments, where an object is moving freely in space surrounding the earth's surface, causes energy to be lost.

In such experiments, for the principle to apply, we would need to create a situation where a vacuum has been created in the region where the object is moving. The air through friction creates a decelerating force that is proportional to the square of the velocity. The energy loss goes into heating the object and the surrounding air molecules. This energy should be included in the equations for exactness and energy is then conserved. The same is true in many other phenomena. In many situations, heat is generated and we do not always know how to handle it. It is very often a complex process. This means that we do not have what is called a cyclic process: if we return to the original situation relative to coordinates, and energy is lost along the path followed, we do not return to the same original conditions. In the equations, something appears to be missing. As we have said previously, we say that the system is not conservative.

Let us reexamine Eq. 2.60, in a closed path when the force is conservative. Then, we come back to the original coordinates in the same state and no work is done around the path:

$$W = \oint \boldsymbol{F} \cdot d\boldsymbol{S} = 0 \tag{2.98}$$

The kinetic energy is the same at the beginning and at the end of the closed-circuit path:

$$W_{12} = T_2 - T_1 = 0 \tag{2.99}$$

If this is valid, the work must be independent of the path followed by the object. We should thus be in a position to express work done in terms of a quantity that depends only on the end points and not on the path followed. We can express that statement mathematically in terms of a quantity reflecting the property of the space the object is in. This quantity is the *potential energy* that we have defined earlier. Using Figure 2.7 we may then write for an infinitesimal path length $d\boldsymbol{S}$ the change in potential energy as:

$$\boldsymbol{F} \cdot d\boldsymbol{S} = -dV \tag{2.100}$$

Thus, we have for a general path, using Eq. 2.60:

$$W_{12} = \int_1^2 -\frac{dV}{dS} \cdot dS = \int_1^2 -dV = -(V_2 - V_1) \tag{2.101}$$

The work to be done is then simply the change in potential energy in going from point 1 to point 2. This is equal to the difference in kinetic energy that we have calculated earlier. Thus, we can write:

$$T_2 - T_1 = V_1 - V_2 \tag{2.102}$$

leading to:

$$T_2 + V_2 = T_1 + V_1 = T + V = constant \tag{2.103}$$

Thus, total energy, that is kinetic and potential energy, is conserved when applied forces are conservative. We have thus proven in an elementary way the principle of *conservation of energy* that we introduced as a postulate.

Conservation of Linear Momentum.

If no force is applied to a material object, its speed does not change. Consequently, if the mass also stays constant, the linear momentum, $\boldsymbol{p} = m\mathbf{v}$, is conserved. This is shown directly from Eq. 2.56:

$$\boldsymbol{F} = \frac{d}{dt}\boldsymbol{p} \tag{2.104}$$

If no force is applied, $\boldsymbol{F} = 0$ and $\boldsymbol{p}$ is a constant of movement independent of time. Thus, the linear momentum is conserved. In collisions between objects, the total linear momentum, that is the sum of all objects' momentum, is also conserved if no external force is applied to the ensemble of objects. A billiard ball rolling on the table sees its speed altered by friction with the surface and the resistance of the air. Its momentum is not conserved. In collisions between two balls in free space, for example, momentum is conserved if energy is not lost as heat inside the balls during collision.

Conservation of Angular Momentum.
We have introduced above the concept of linear momentum. We have also seen that it is at the basis of Newton's second principle: when a system is conservative, that is to say the force is obtained from the gradient of a potential, then that entity is conserved. There is, however, another entity that plays a very important role in describing the functioning of the universe. This is the *angular momentum* $\boldsymbol{\ell}$. It is defined as the moment of the linear momentum or the vector product of $\boldsymbol{r}$ and $m\mathbf{v}$:

$$\boldsymbol{\ell} = \boldsymbol{r} \times m\mathbf{v} \tag{2.105}$$

with the vectors $\boldsymbol{r}$ and $\mathbf{v}$ defined by means of Figure 2.10:

We may introduce also the moment of an added force F that could be acting on mass m and change its velocity and direction. The entity is called torque $\boldsymbol{N}$ and is represented by means of the vector product of $\boldsymbol{r}$ and $\boldsymbol{F}$:

$$\boldsymbol{N} = \boldsymbol{r} \times \boldsymbol{F} \tag{2.106}$$

$\boldsymbol{F}$ may be written by means of Newton's second principle (Eq. 2.56). We obtain:

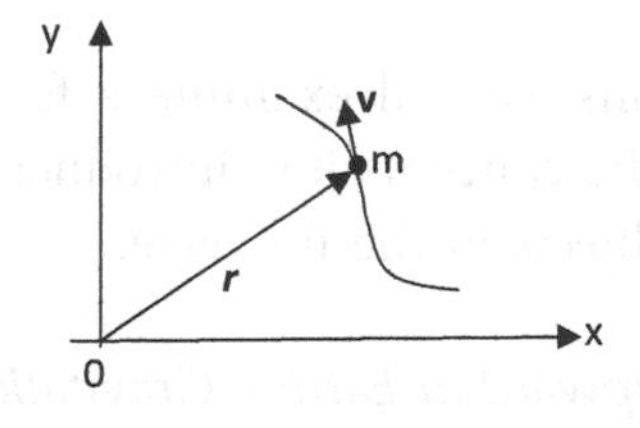

FIGURE 2.10 Representation of the vectors entering into the definition of the angular momentum.

$$N = r \times \frac{d}{dt}(m\mathbf{v}) \tag{2.107}$$

This may be written differently. The derivative of the vector product $r \times m\mathbf{v}$ may be written as:

$$\frac{d}{dt}(r \times m\mathbf{v}) = \frac{dr}{dt} \times m\mathbf{v} + r \times \frac{d}{dt} m\mathbf{v} \tag{2.108}$$

Since dr/dt is equal to **v** and $\mathbf{v} \times \mathbf{v}$ involves the sine of the angle between two identical vectors, the first term on the right hand side is zero and Eq. 2.107 becomes:

$$N = \frac{d}{dt}(r \times m\mathbf{v}) \tag{2.109}$$

This is simply the derivative of the angular momentum defined above by means of Eq. 2.105

$$N = \frac{d}{dt}\ell \tag{2.110}$$

This is very similar to Eq. 2.104 for the linear momentum and it has an important property. We note immediately that if the torque N is zero, the angular momentum ℓ stays constant. This is essentially an elementary proof of the *conservation of angular momentum* when no torque is applied.

Illustrative Examples

In the following sections we will examine a few elementary examples illustrating the use of the concepts just introduced to analyze the actual dynamics of selected objects in the universe.

Object Sent Directly Upward in Earth's Gravitational Field

We can look at an example of the principle of conservation of energy by making an experiment in the gravitational environment near the surface of the earth. Let us look at the behavior of an object of mass m that we send straight up at speed v_o from the surface of the earth. As shown in Figure 2.11 it will go up fighting against the gravitational attraction of the earth. It is thus decelerated by that force, reaches a certain height and falls back to its point of origin.

We can calculate the behavior of the object on its path up and down. Its speed is v_o at time t=0 at height h = 0. In going up its velocity is reduced by the opposite acceleration (deceleration) due to gravity. We assume that the object stays close to the earth's surface. We thus use g as a constant within the approximation mentioned earlier. The speed reduction with time is -gt and we thus have:

$$v_h = v_o - gt \tag{2.111}$$

The height reached after a time t is given by

$$h = \int v_h dt = \int v_o dt - \int gt dt \tag{2.112}$$

or

$$h = v_o t - \frac{1}{2} g t^2 \tag{2.113}$$

Making t explicit from Eq. 2.111, and replacing it in the last equation we obtain:

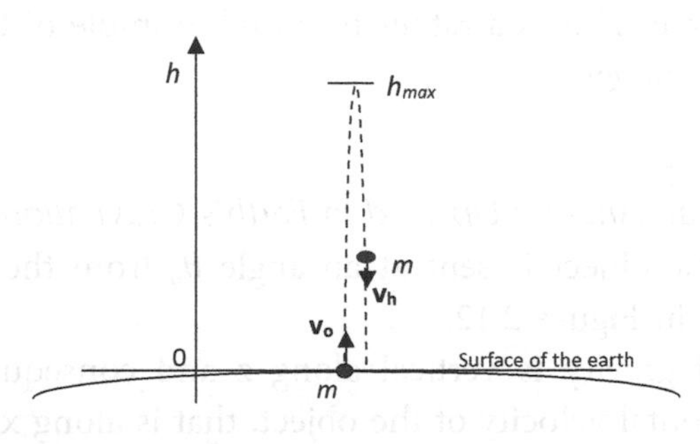

FIGURE 2.11 Motion of an object of mass m thrown up in the gravitational field of the earth.

$$v_h^2 = v_o^2 - 2gh \tag{2.114}$$

We can then multiply this last equation by $(1/2)m$ and obtain:

$$\frac{1}{2}mv_h^2 = \frac{1}{2}mv_o^2 - mgh \tag{2.115}$$

Since we are only interested in differences we have displaced the origin for the potential energy to the surface of the earth as we have done earlier. The term mgh is thus the potential energy of the object at height h. We see immediately that the kinetic energy of the object, which was $\frac{1}{2}mv_o^2$ at height $h=0$, is continuously reduced by the quantity mgh on its way up. When the height reached is such that $mgh = (1/2)mv_o^2$, its speed v_h is 0 and maximum height h_{max} is reached. The object then falls back and returns to its origin h=0 with a velocity v_o. During the whole travel, the relation

$$\frac{1}{2}mv_h^2 + mgh = \frac{1}{2}mv_o^2 \tag{2.116}$$

holds and we see that the total energy of the object at the beginning remains constant, the kinetic energy being transformed into potential energy on its way up. When the object falls back, the potential energy is recovered as kinetic energy. The final speed at the time the object hits the ground is v_0. Energy is fully conserved on the way up and down. It is only exchanged between potential and kinetic energy. Gravity is a conservative force. This is a rather beautiful example of the principle of *conservation of energy.*

Object Sent at an Angle θ Upward in Earth's Gravitational Field

Suppose that the object is sent at an angle θ_o from the surface of the earth as shown in Figure 2.12.

The force of gravity is vertical along z and consequently does not affect the horizontal velocity of the object, that is along x. At the origin, the velocities are given by:

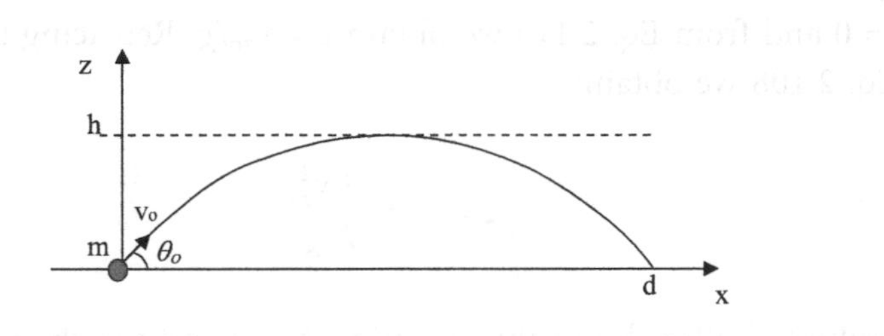

FIGURE 2.12 Trajectory of an object of mass *m* sent at an angle θ_o from the surface of the earth.

$$v_{xo} = v_o \cos\theta_o \quad v_{zo} = v_o \sin\theta_o \tag{2.117}$$

Since the horizontal velocity is not altered, the distance travelled along axis x is given by:

$$x = v_{xo}t = v_o t \cos\theta_o \tag{2.118}$$

On the other hand, the velocity v_z upward is affected by gravity and is given by:

$$v_z = v_{zo} - gt \tag{2.119}$$

Integrating this equation over time, we obtain *z* as:

$$z = v_{zo}t - \frac{1}{2}gt^2 \tag{2.120}$$

Using Eq. 2.118 for t and replacing it in this last equation we have:

$$z = \frac{v_{zo}}{v_{xo}}x - \frac{1}{2v_{xo}}gx^2 \tag{2.121}$$

The trajectory is a parabola. The maximum can be calculated by taking the derivative of *z* and equating it to zero. When the object reaches h, the maximum height of the trajectory, we have

$v_z = 0$ and from Eq. 2.119 we obtain $t = v_{zo}/g$. Replacing this value of t in Eq. 2.108 we obtain:

$$z = h = \frac{1}{2}\frac{v_{zo}^2}{g} \tag{2.122}$$

We multiply both sides of this equation by m and we obtain:

$$mgh = \frac{1}{2}mv_{zo}^2 \tag{2.123}$$

showing that the maximum height is reached when the vertical kinetic energy is totally transformed into potential energy. This validates again the principle of conservation of energy that we could have used to make the analysis and obtain the result above.

The Simple Pendulum

Assume a ball of mass m attached to a string of length l, itself attached to a fixed point such as a ceiling. We have a pendulum as shown in Figure 2.13.

We set the zero of the system of axis (x, y) at the junction of the string to the ceiling. The mass m is submitted to the gravitational attraction of the earth, resulting in a force oriented downward and equal to mg. We restrain the motion of the mass to the x–y plane by

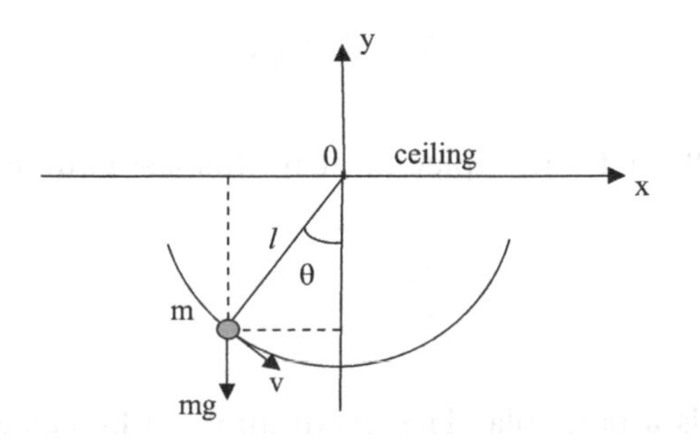

FIGURE 2.13 Representation of the motion of a mass m in an x–y plane with one constraint: the mass is attached to the ceiling by means of a string of length l.

means of a condition fixed when the system is first set in motion. This is a first constraint and, consequently, the variable z does not appear in the configuration space. The length of the vector $\boldsymbol{r}$, determining the trajectory of the ball, is fixed and does not change with time. It is the arc of a circle of radius l and we necessarily have:

$$x^2 + y^2 - l^2 = 0 \quad \text{or} \quad |r| = l = \sqrt{x^2 + y^2} \tag{2.124}$$

This is the equation of the second constraint and consequently x and y are not independent. There are thus two constraints on the motion of the ball and there is only one degree of freedom. It is actually the angle θ. We have thus passed from a Cartesian system of three coordinates (x, y, z) to a system of one coordinate θ, which determines entirely the evolution of our system in configuration space. θ is our generalized coordinate. From earlier analysis, we have a relation between θ and x and y. Since only one coordinate or degree of freedom is involved, the analysis is considerably reduced. In fact, we have:

$$y = l\cos\theta \qquad x = l\sin\theta \tag{2.125}$$

Consequently, if one coordinate, say x, is set, then θ is set and the other coordinate is also set, l being a constant.

The following calculation illustrates well the use of generalized coordinates. The force exerted on mass m by gravity is mg and the potential energy depends on the angle θ that varies from zero to a maximum on each side of the vertical. We set the point 0 of potential energy at the minimum height when θ is 0. The height h of the ball above that point is l(1-cosθ). Thus, the gravitational potential energy of mass m is given by:

$$V = mg(1 - \cos\theta) \tag{2.126}$$

On the other hand, the speed of the ball ls given by:

$$\mathrm{v} = l\dot{\theta} \tag{2.127}$$

$\dot{\theta}$ being maximum when θ is zero and 0 when θ is maximum. The kinetic energy of the ball is thus given by:

$$T = \frac{1}{2} ml^2 \dot{\theta}^2 \tag{2.128}$$

Since it is assumed that there is no friction we can use the principle of conservation of energy demonstrated earlier and write:

$$T + V = \frac{1}{2} ml^2 \dot{\theta}^2 + mgl(1 - \cos\theta) = constant = E_p \tag{2.129}$$

Let us call θ_m the angle of maximum deflection of the pendulum from the vertical. At that angle motion is reversed and $\dot{\theta}$ is zero. We thus have:

$$mgl(1 - \cos\theta_m) = E_p \tag{2.130}$$

We combine this result with Eq. 2.129 and obtain:

$$\dot{\theta}^2 + 2\frac{g}{l}(cos\theta_m - \cos\theta) = 0 \tag{2.131}$$

For small amplitudes of oscillation, we can approximate the cosine terms by $(1 - \theta^2)$ resulting in:

$$\dot{\theta} = \frac{d\theta}{dt} = \sqrt{\frac{g}{l}(\theta_m^2 - \theta^2)} \tag{2.132}$$

This equation can be integrated easily and the result is:

$$\theta = \theta_m \sin(\sqrt{\frac{g}{l}}\ t) \tag{2.133}$$

in which the constant of integration has been set equal to zero by fixing the origin of time (0) at position $\theta = 0$. The final result is a harmonic motion of frequency $\omega = \sqrt{\frac{g}{l}}$. This is a rather interesting example of the use of generalized coordinates in establishing the dynamics of an object attached by a string to a fixed point, using simple algebra.

Orbital Motion

As another example of the use of Newton's principle, let us calculate in an elementary way the motion of a small mass m orbiting a large mass M as in Figure 2.14.

The orbit is in the x–y plane and consequently the angle θ is $\pi/2$. We show the motion of mass m in greater detail in Figure 2.15.

We will do a simple analysis of the dynamics of such a system using the principles just enunciated above. We will arrive at a solution that allows stable orbits of the small object with mass m around the larger mass M and we will derive the condition for a circular orbit. We will see that the algebra using vector analysis is a little involved. In a later chapter we will use Lagrange's approach that, although not introducing new physics, arrives at the same results more easily.

In principle, if the two masses were similar, they would form a complex system, both objects revolving around a point in space determined by the relative mass of the bodies and called the center of mass. In the present case, we simply assume that mass M is much larger than the small one, m. Mass M essentially does not move and it is considered at rest in the system of coordinates chosen. In such a case, we can assume that the system of coordinates is centered on the large mass and does not move. Let us represent the motion of the small mass m in

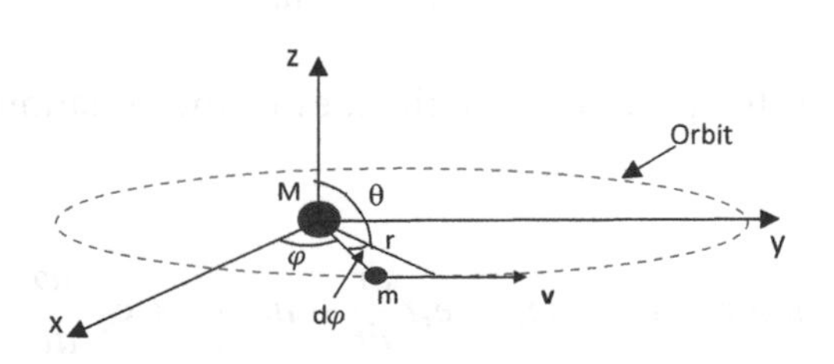

FIGURE 2.14 Orbital motion of an object of mass m around another object of much greater mass M. The coordinate system is fixed with its origin at the center of mass M. For simplicity the orbit is assumed to be in the x–y plane.

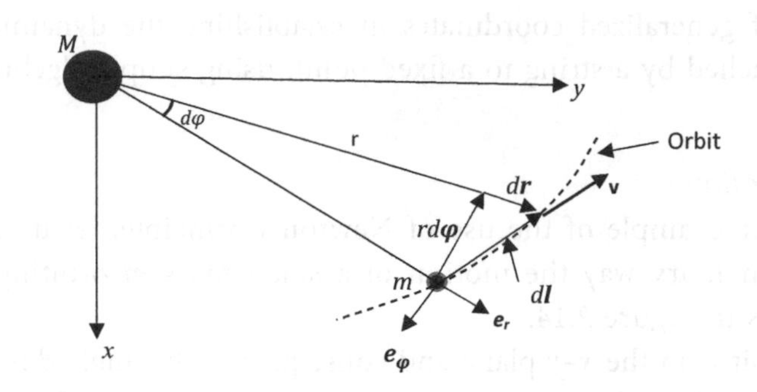

FIGURE 2.15 Representation of various vectors in the x–y plane required in the analysis of the motion of mass m around the larger mass M.

vector notation as in Figure 2.15. From the figure we write the speed of the object $\mathbf{v}$ as:

$$\mathbf{v} = \frac{d\mathbf{l}}{dt} \tag{2.134}$$

From the same figure, we also have:

$$d\mathbf{l} = \mathbf{e_r}dr + \mathbf{e_\varphi}rd\varphi \tag{2.135}$$

Thus, we have

$$\mathbf{v} = \mathbf{e_r}\frac{dr}{dt} + \mathbf{e_\varphi}\frac{rd\varphi}{dt} \tag{2.136}$$

We make the vector product of both sides of the equation by $m\mathbf{e_r}r$. We obtain:

$$m\mathbf{e_r}r \times \mathbf{v} = m\mathbf{e_r} \times \mathbf{e_r}r\frac{dr}{dt} + mr^2\mathbf{e_r} \times \mathbf{e_\varphi}\frac{d\varphi}{dt} \tag{2.137}$$

A vector product involves a multiplication by the sine of the angle between the two vectors in question. Consequently, $\mathbf{e_r} \times \mathbf{e_r}$ is equal to zero. The second term on the right hand side consists of the vector product of two orthogonal unit vectors $\mathbf{e_r}$ and $\mathbf{e_\varphi}$. It produces a unit vector $\mathbf{e_{r\varphi}}$ perpendicular to the plane formed by the two vectors:

$$mr \times \mathbf{v} = mr^2 \mathbf{e_{r\varphi}} \frac{d\varphi}{dt} = mr^2 \mathbf{e_{r\varphi}} \dot{\varphi} = \ell \tag{2.138}$$

The resulting vector $mr \times \mathbf{v}$ is the moment of momentum $r \times m\mathbf{v} = r \times p$ and was defined earlier as the angular momentum, ℓ. The angular speed $\dot{\varphi}$ can thus be expressed as:

$$\dot{\varphi} = \frac{\ell}{mr^2} \tag{2.139}$$

and ℓ is thus an indication of the angular speed.

An important aspect of the orbital motion is the energy of the object of mass m in motion. It is composed of its kinetic energy and its potential energy in the gravitational field of the larger mass M. The gravitational potential energy is given by Eq. 2.83 while the kinetic energy is given by Eq. 2.62. The square of the velocity is obtained from the scalar product of $\mathbf{v}$ by itself. We thus have:

$$v^2 = \mathbf{v} \cdot \mathbf{v} = (\mathbf{e_r} \frac{dr}{dt} + \mathbf{e_r} \frac{rd\varphi}{dt}) \cdot (\mathbf{e_\varphi} \frac{dr}{dt} + \mathbf{e_{r\varphi}} \frac{rd\varphi}{dt}) = \dot{r}^2 + r^2 \dot{\varphi}^2 \tag{2.140}$$

The total energy of the object is thus:

$$E_{total} = \frac{1}{2} m (\dot{r}^2 + r^2 \dot{\varphi}^2) - G \frac{Mm}{r} \tag{2.141}$$

and using Eq. 2.134, we obtain:

$$E_{total} = \frac{1}{2} m \left(\dot{r}^2 + \frac{\ell^2}{m^2 r^2} \right) - G \frac{Mm}{r} \tag{2.142}$$

or

$$E_{total} = \frac{1}{2}m\dot{r}^2 + \left(\frac{\ell^2}{2mr^2} - G\frac{Mm}{r}\right) \tag{2.143}$$

This equation can be interpreted in a very interesting way. It looks like the equation of a single object having kinetic energy $\frac{1}{2}m\dot{r}^2$ in a potential V_{eff} given by:

$$V_{eff} = \left(\frac{\ell^2}{2mr^2} - G\frac{Mm}{r}\right) \tag{2.144}$$

The total energy is thus:

$$E_{total} = \frac{1}{2}m\dot{r}^2 + V_{eff} \tag{2.145}$$

This exercise leads effectively to a one-dimensional problem with $\boldsymbol{r}$ as variable V_{eff} being dependent only on coordinate $\boldsymbol{r}$. To analyze the behavior of the system we draw this effective potential to make explicit the well (exaggerated) that is created by the interaction as a function of r. This is shown in Figure 2.16.

We shall examine a few situations by setting the value of the total energy of a mass m in orbit around a much larger mass M. From the principle introduced earlier we know that the total energy E is conserved around the orbit. Objects in motion around mass M see their position $\boldsymbol{r}$ vary but their energy is constant. Consequently, there is a continual exchange between potential and kinetic energy in the orbit. Referring to Figure 2.16, assume first that the total energy is E_3. The object comes from a large distance where its energy is mostly kinetic and originates from a change in distance $\boldsymbol{r}$ in the system of coordinates fixed to the large mass M. In fact, according to Eq. 2.145, the radial kinetic energy of the object is given by the difference between E_3 and V_{eff}, $\frac{1}{2}m\dot{r}^2$. When the object reaches distance r_3 its radial velocity is zero. It cannot go any closer to the large mass because its radial kinetic energy would need to be negative giving rise to an imaginary velocity, which is not possible.

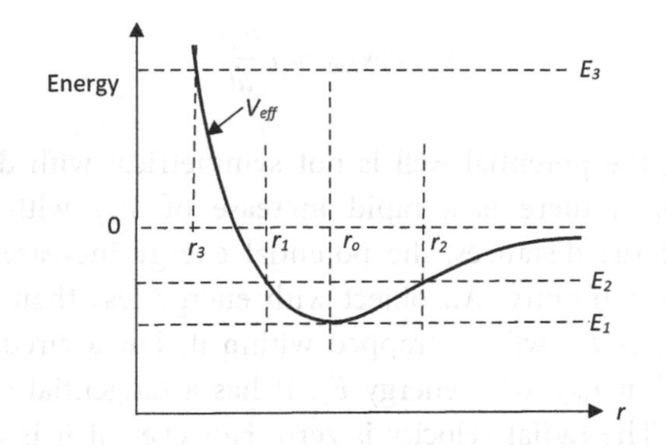

FIGURE 2.16 V_{eff} drawn as to make evident the potential well created by the interaction between a small mass m and a large mass M, taking into account gravity and relative motion. r is the distance between the two masses.

The object essentially hits the barrier of V_{eff}, which in common language can be seen as a repulsive centrifugal barrier.

The angular momentum is large, and the object is deviated by gravity into an unbounded motion back to infinity. On the other hand, if the total energy is E_1 we see that this energy is possible only within the potential well. The system sits at the bottom of the well, which is an extremum. We can find that position by differentiating V_{eff} with respect to r and equate the result to zero. We obtain:

$$\frac{d}{dr}V_{eff} = -\frac{\ell}{mr^3} + \frac{GMm}{r^2} = 0 \qquad (2.146)$$

We use Eq. 2.139 for the angular momentum and find that the minimum arises at:

$$r_{min} = r_o = \frac{GM}{\mathrm{v}_\varphi^2} \qquad (2.147)$$

where we have defined the angular velocity v_φ for the circular orbit as the tangential velocity:

$$v_\varphi = v_{tan} = r\frac{d\varphi}{dt} \tag{2.148}$$

We note that the potential well is not symmetrical with distance r. At low values of r, there is a rapid increase of V_{eff} with diminishing distance. At large distances, the potential energy increases slowly and tends to zero at infinity. An object with energy less than the depth of the well, such as E_2, will be trapped within it. For a circular orbit, the object is fixed at $r_{\min}$ with energy E_1. It has a tangential velocity given by Eq. 2.148. The radial velocity is zero. However, if it is not at r_o and has an angular momentum compatible with the effective potential shown in Figure 2.16, it may simply oscillate within the well with $\boldsymbol{r}$ changing from r_1 to r_2 in a cyclic way. This situation is that of an object orbiting the large mass, but not in a circular orbit. The value of r changes and the velocity of the object will be composed of both $\dot{r}$ and $\dot{\varphi}$ components. The object may describe elliptical orbits, with the large mass at one of the focal points of the ellipse.

The reader is invited to do an exercise using the actual physical parameters given at the end of this text for the sun-earth system. It looks very much like a single ensemble in a potential well with a minimum energy at $r_{\min}$. If the earth is situated at that distance and its tangential speed is appropriate, it will keep going around the sun without inhibition in its motion and will keep its speed and distance permanently unless something not taken into account in the analysis happens. If we set $r_{\min}$ as the actual measured average radius of the orbit of the earth around the sun (1.50×10^{11} m) we can calculate from Eq. 2.147 the average velocity required for the system to be at that minimum potential energy. It is 29.8 km/s, which is the velocity observed by means of measurements.

C. RELATIVITY

Moving Reference Frames

At this point we need to introduce the important concept of the relativity of movement. In fact, such a concept was discovered in the early days of the developments of principles introduced above. It was quickly realized by scientists in the early development of physics that it is not possible to determine the value of the constant velocity of an object in space unless a point of reference is provided. And even then, it is not possible to say what is moving: the point of reference provided or the object in question. We experience that in an evident way when, in a train on a railway station, we have the sensation of moving when a wagon on another track starts to move relative to us. We have to look through the window in another direction in order to determine the actual situation. Only relative velocity can be determined, and velocity is entirely connected to the frame of reference we are using. If the frame of reference in which we claim we are at rest moves at the same constant velocity as the object in motion, that object appears to us to be at rest and we have to consider it in a state of rest. What happens to the laws of mechanics in systems in relative motion? We can represent the situation by means of Figure 2.17 in which we show two systems of reference Σ and Σ' in motion relative to each other.

To simplify matters let us assume 2D systems, with Σ' moving at velocity $\mathbf{v}$ along the x axis of system Σ, both axes x and x′ coinciding. Assume an object with mass m at position $\boldsymbol{r}$ in system Σ and $\boldsymbol{r}'$ in system Σ'. From the figure, we have:

$$\boldsymbol{r} = \mathbf{v}t + \boldsymbol{r}' \tag{2.149}$$

The velocity of m in $\sum'$ is obtained directly by differentiation of $\boldsymbol{r}'$. We assume $\mathbf{v}$ is constant and we have:

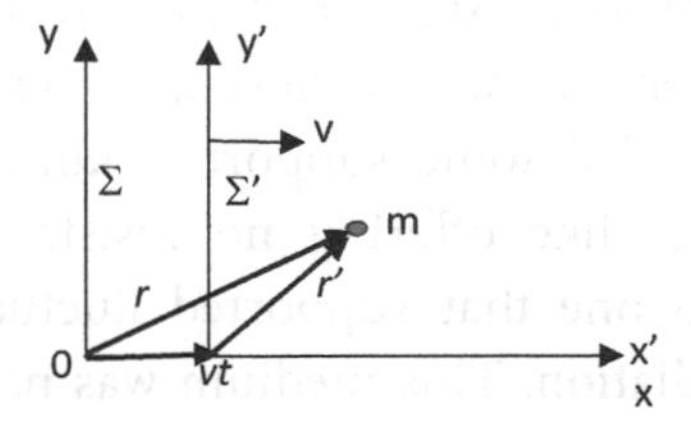

FIGURE 2.17 Two systems of coordinates in motion at velocity v relative to each other.

$$\dot{\boldsymbol{r}} = \mathbf{v} + \dot{\boldsymbol{r}}' \tag{2.150}$$

If a force is applied to m, acceleration results. This acceleration $\boldsymbol{a}$ is obtained by means of a second differentiation and since **v** is constant we have:

$$\ddot{\boldsymbol{r}} = \ddot{\boldsymbol{r}}' = \boldsymbol{a} \tag{2.151}$$

Consequently, acceleration in both systems is the same and Newton's second law, $\boldsymbol{F}=m\boldsymbol{a}$, is also the same in both systems. In Eq. 2.150, **v** can be moved to the left-hand side and system Σ may be considered as the one moving away from system Σ' in the negative direction. Eq. 2.151 still applies. The same conclusion is reached. Consequently, in that context, it is assumed that the mechanical laws of physics are the same in both systems and it is not possible to tell which system is moving. Both systems are inertial systems. There is no way for an observer fixed to a system in uniform motion relative to another system to tell if he is moving or if the other is the one moving. There is no absolute reference point.

The kind of transformation just described from one system to another in relative motion at velocity **v** is called a Galilean transformation.

Special Relativity and Lorentz Transformation

We may wonder, however, if we could not find a situation in which a point of reference or a phenomenon, which, when observed appropriately, could be used as an absolute reference frame, that is a frame relative to which all matters in the universe could be referred to. We would then have an absolute space in which all positions and motions could be set absolutely. It was believed, some time ago, that light travelling in space needed some support, a kind of medium that had very special properties like offering no resistance to the motion of objects in it, but also one that supported fluctuations of electric and magnetic fields, or radiation. This medium was named "*ether*". It would operate similarly to the role air plays in the transmission of sound. This medium would offer a kind of grid providing an absolute reference to which motion could be referred. Stars and planets would travel freely through it. On the other hand, it was found that light travels at speed c, approximately 300,000 km per second. Consequently, it was concluded that if an observer travels at a velocity $\mathbf{v}$ in that ether, opposite to the direction of propagation of a pulse of electromagnetic radiation, he would observe that the pulse is passing by at a velocity greater than c. He would then be able to tell that he is moving relative to the ether. This hypothesis was tested by means of the motion of the earth travelling at about 30 km per second in its orbit around the sun. This was tested at the end of the 19th century by Michelson and Morley, who used an interferometer that bears their name. It was found that there is no observable effect that would be compatible with the presence of a supporting medium, the ether, the presumed support for the transmission of electromagnetic radiation. Several other precise tests were made later which confirmed this result. It was concluded that the speed of light is c and, when measured, is not altered either by the velocity of the source or of the observer. This result is truly fascinating. It looks very much like a conspiracy of nature to prevent the measurement of motion in an absolute way. Actually, this property was already imbedded in Maxwell's equations. These equations connect in an implicit way magnetic and electric fields by means of their dependence on time. They say that if a magnetic field is varying in time it creates an electric field. The reverse is also true. If an electric field changes over time a magnetic field is created. As we will see below, Maxwell's equation then implies that those fluctuations propagate through space without restraint at speed c, without reference to any coordinate system. The propagation obeys a differential equation, called the wave equation. For the electric field it is:

$$\nabla^2 \boldsymbol{E} = \frac{1}{c^2}\frac{\partial^2 \boldsymbol{E}}{\partial t^2} \tag{2.152}$$

with a similar equation for the magnetic field $\boldsymbol{B}$. In that equation, c is the constant of propagation, the speed of light. In view of the conclusion reached above relative to that constant, it is clear that we cannot apply Galileo's transformation to transform that equation in passing from one inertial system of reference to another one in relative motion. Einstein took a great step towards resolving this question through the introduction of two basic postulates:

- The speed of light c is a constant in all reference frames, independent of the velocity of the source of radiation or of the observer.
- All laws of physics are the same in all inertial systems (not accelerated). This postulate essentially reflects a symmetry property of the laws of physics in space.

All mechanical and electromagnetic laws apply in inertial systems independent of their relative velocity. For example, Eq. 2.152 applies in two systems moving relative to each other at constant velocity. It is thus clear that Eq. 2.152 cannot be transformed to a system moving at velocity **v** relative to an observer at rest by adding **v** to the speed c. Consequently, the Galilean transformation that we have used above cannot be correct for transforming the radiation propagation law from one system in relative motion to another one considered at rest. We need to introduce a new transformation that will keep Eq. 2.152 the same in all inertial systems whatever their relative velocities. This transformation is called the Lorentz transformation. It is not clear how such a transformation will alter Newton's mechanical laws in passing from one inertial system to another one in relative motion. However, we know that Newton's laws are valid under Galilean transformation at low velocities. Thus, it must be concluded that Newton's laws are approximations of relativistic laws that obey the relativity postulates.

In order to derive the Lorentz transformation, we will introduce the concept of spacetime in which time is added as a coordinate to the standard x, y, z coordinates of a Cartesian system, essentially forming a four-dimensional spacetime system. There are several ways that the transformation can be derived. Many texts have been written on the subject. We give here a summary of a derivation presented in Goldstein (*Classical Mechanics* 1959). That type of analysis shows that the passing from one frame of reference considered at rest to one in motion at velocity **v** corresponds essentially to a rotation of axes in that four-dimensional system.

Let us situate ourselves in a coordinate system at rest that we will call Σ. Let us introduce another system called Σ' in constant linear motion along the z axis at velocity **v** and coinciding with Σ at time t=0. This is shown in Figure 2.18, using Cartesian coordinates *x, y* and *z*.

Let us fix at the origin of Σ a source that emits a pulse of radiation at time *t*=0. This radiation spreads in all directions at speed *c*. The wave front travels on a circle centered on the origin *x*=*y*=*z*=*0*. The equation describing the evolution of this wave front is:

$$x^2 + y^2 + z^2 = c^2t^2 \tag{2.153}$$

The right-hand member of the equation, *ct* squared, identifies the role of the radius of the sphere increasing with the wave front. We assume that at time *t*=0 both systems Σ and Σ′ coincide. However, system Σ′ is in motion at velocity **v** along z in frame Σ. Should we take into account that velocity in the propagation of the wave in system Σ′? Is velocity **v** adding vectorially to the speed *c* of the wave front? If it were the case, velocity **v**, being unidirectional, would affect the spherical symmetry of the wave front as seen in system Σ′. This is not observed in practice and the wave propagate in both systems at the same speed *c* as was described in the main text. As a postulate, *c* is considered to be a universal constant and light propagates at that speed independently of the velocity of the source or of the observer. Consequently, Eq. 2.153 is a valid equation in all reference frames. We can thus write in frame Σ′:

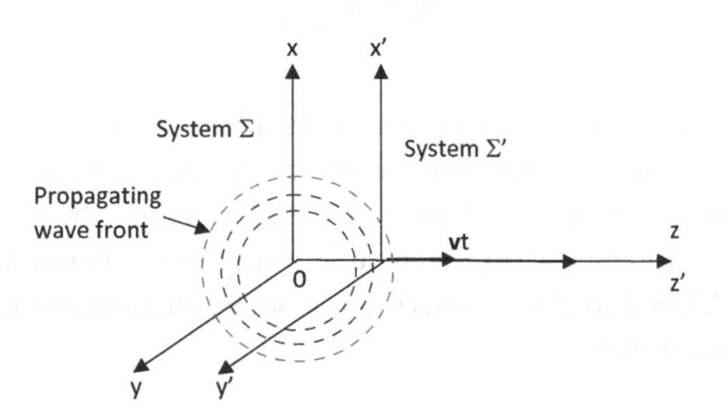

FIGURE 2.18 Coordinate system Σ′ in uniform motion at speed v along axis z relative to a coordinate system Σ.

$$x'^2 + y'^2 + z'^2 = c^2 t'^2 \quad (2.154)$$

We have made the wave propagation speed c the same as in system Σ. However, we have allowed t to be different. We note that x and y stay the same in both systems, but position z of frame Σ' varies with time as:

$$z = \mathrm{v}t \quad (2.155)$$

To simplify notation, we will use a 4-D representation in which ct is a fourth coordinate. In that case, as is done generally in this context, symbol x_i with i=1,2,3,4 is used to represent coordinates, x, y, z and ct. We can thus write Eq. 2.153 as:

$$\sum x_i^2 = 0 \quad (2.156)$$

with the understanding that on the left side we have a summation in which $x_4^2 = c^2 t^2$ is a negative term. This 4-D space is spacetime; it is Minkowski space; it is world-space. We can interpret it as if x_4 is an imaginary term ict. We can also write a similar expression for the wave observed in system Σ' in motion at speed v. The speed of light stays the same in that system and Eq. 2.156 can be written using the same notation:

$$\sum x_i'^2 = 0 \quad (2.157)$$

Since the speed of light is assumed to be identical in both systems and only the axes are changed, the observers in each system see the same thing: a wave propagating at speed c. Consequently, the moving observer's system of axes will adjust in such a way that c is not altered. Both equations 2.156 and 2.157 describe the same phenomenon. We equate them and we write:

$$\sum \mathrm{x}_\mathrm{i}'^2 = \sum \mathrm{x}_\mathrm{i}^2 \text{ where } \mathrm{i} = 1,\ 2,\ 3,\ 4 \quad (2.158)$$

We may then assume a unique relation between the coordinates as we would do with 2D and 3D systems in rotation relative to each other, as was done earlier. We thus write:

$$x_i' = \sum_{j=1}^{4} a_{ij} x_j \tag{2.159}$$

where the a_{ij} are transformation parameters connecting the coordinates. In Figure 2.18, frame Σ' moves at velocity v. It traverses in frame Σ a distance $\mathbf{v}t$ as given by Eq. 2.155. In frame Σ everything seems at rest except for the wave front moving along all three axes at speed c. On the other hand, distances along x_1 and x_2 stay the same in both systems. This is fundamental since they are not at all involved with the displacement along x_3. Any object situated along these axes does not move except in the x_3 direction. Thus $x_1 = x'_1$ and $x_2 = x'_2$.

The relation between the coordinates from Σ to Σ' may be written in a matrix form:

$$[x'] = [A]\,[x] \qquad (\textit{matrix multiplication}) \tag{2.160}$$

The transformation from one system to the other is thus done with a transformation matrix A with elements affecting x_3 and x_4 or x_3' and x_4' (z, z', t and t') only:

$$A = \begin{vmatrix} 1 & 0 & 0 & 0 \\ 0 & 1 & 0 & 0 \\ 0 & 0 & a_{33} & a_{34} \\ 0 & 0 & a_{43} & a_{44} \end{vmatrix} \tag{2.161}$$

We need to calculate the a_{ij} elements. We detail the operation implicit in the matrix multiplication in Eq. 2.160. Due to the orthogonality of the axes x_i, the elements must fulfill relations similar to those of Eqs. 2.26 and 2.27. Thus, we have:

$$a_{33}^2 + a_{34}^2 = 1 \tag{2.162}$$

$$a_{43}^2 + a_{44}^2 = 1 \tag{2.163}$$

$$a_{33}a_{43} + a_{34}a_{44} = 0 \tag{2.164}$$

We need to make explicit the relations between coordinates space x_3, time x_4 and coordinates x_3' and time x_4'. We first have in Σ:

$$x_3 = \text{v}t \tag{2.165}$$

This can be written:

$$\text{v}t = \frac{\text{v}}{c}ct \tag{2.166}$$

But $x_4 = ict$ and the third and fourth coordinate x_3 and x_4 are coupled. This is the main difference from the 3-D system we examined earlier in which all coordinates were independent. Thus, we can write:

$$x_3 = -i\beta x_4 \tag{2.167}$$

where β is equal to v/c. However, according to Eqs. 2.160 and 2.161, we also have:

$$x_3' = a_{33}x_3 + a_{34}x_4 \tag{2.168}$$

Using Eq. 2.167 we thus have:

$$x_3' = x_4(a_{34} - i\beta a_{33}) \tag{2.169}$$

However, this coordinate x_3' in Σ′ is the origin 0 since x_3 in Σ is vt. Consequently, Eq. 2.169 is equal to 0. Since x_4, the time coordinate cannot be 0, we need:

$$(a_{34} - i\beta a_{33}) = 0 \tag{2.170}$$

or

$$a_{34} = i\beta a_{33} \tag{2.171}$$

Replace in the orthogonality Eq. 2.162 and we obtain:

$$a_{33}^2\left(1-\beta^2\right) = 1 \tag{2.172}$$

or

$$a_{33} = \frac{1}{\sqrt{1-\beta^2}} \tag{2.173}$$

and from Eq. 2.171:

$$a_{34} = \frac{i\beta}{\sqrt{1-\beta^2}} \tag{2.174}$$

The other elements of the matrix can be obtained similarly from the other orthogonality equations.

From Eq. 2.164 we have

$$a_{43} = -a_{44}\frac{a_{34}}{a_{33}} = -i\beta a_{44} \tag{2.175}$$

which using Eq. 2.163 gives:

$$a_{44} = \frac{1}{\sqrt{1-\beta^2}} \tag{2.176}$$

The last matrix element is then:

$$a_{43} = \frac{-i\beta}{\sqrt{1-\beta^2}} \tag{2.177}$$

We thus have all elements of the transformation matrix from a rest frame to a frame moving at velocity v in a 4D space with *x, y, z, ct* coordinates. Matrix 2.161 is thus:

$$A = \begin{vmatrix} 1 & 0 & 0 & 0 \\ 0 & 1 & 0 & 0 \\ 0 & 0 & \frac{1}{\sqrt{1-\beta^2}} & \frac{i\beta}{\sqrt{1-\beta^2}} \\ 0 & 0 & \frac{-i\beta}{\sqrt{1-\beta^2}} & \frac{1}{\sqrt{1-\beta^2}} \end{vmatrix} \tag{2.178}$$

We note that the four elements in the lower block look very much like a rotation in a 2-D system whose elements we have calculated previously and are given in Eq. 2.38. We can evaluate the angle φ of that rotation as follows. Using Eq. 2.38, we may write

$$cos\varphi = \frac{1}{\sqrt{1-\beta^2}} \tag{2.179}$$

Since the right-hand member is larger than 1, the angle cannot be real. If we assume it is imaginary we have

$$\cos i\varphi = ch\,\varphi = \frac{1}{2}(e^{\varphi} + e^{-\varphi}) \tag{2.180}$$

where $ch\ \phi$ is a hyperbolic cosine and can be larger than 1. Since the time axis is essentially an imaginary axis *ict*, this is not surprising. The introduction of a time axis on which our coordinate can only increase and on which we cannot go backward is rather special.

However, using matrix A, we can immediately write the relations between the two systems of coordinates. We use the Cartesian notation *x, y, z, ict* to make things more explicit. Recall that the transformation is achieved by means of Eq. 2.160, which means the multiplication of a 1x4 matrix (representing a one-column vertical matrix with components *x, y, z, ict*) by matrix *A* given by Eq. 2.178, a four-column by four-row matrix. Recall that β is v/c and we obtain:

$$x' = x \tag{2.181}$$

$$y' = y \tag{2.182}$$

$$z' = \frac{z - \mathrm{v}t}{\sqrt{1 - \beta^2}} \tag{2.183}$$

$$t' = \frac{t - \mathrm{v}z/c^2}{\sqrt{1 - \beta^2}} \tag{2.184}$$

These are the Lorentz transformations between 3-D relativistic coordinates systems in relative motion along the z axis. The only assumption we have made is that the speed of light stays the same in any systems moving at velocity **v** relative to each other. Note that the following expressions are also valid in the Σ system since it can be considered moving at velocity –**v** relative to Σ′.

$$x = x' \tag{2.185}$$

$$y = y' \tag{2.186}$$

$$z = \frac{z' + \mathrm{v}t'}{\sqrt{1 - \beta^2}} \tag{2.187}$$

$$t = \frac{t' + \mathrm{v}z'/c^2}{\sqrt{1 - \beta^2}} \tag{2.188}$$

Suppose that in the rest frame Σ we have a rod of length $L = z_2 - z_1$. A traveler passes by at velocity **v** and in his system measures the rod by locating the two end points z_1' and z_2' at the same time t'. From Eq. 2.187, his measurements give:

$$z_1 = \frac{z'_1 + vt'}{\sqrt{1-\beta^2}} \tag{2.189}$$

$$z_2 = \frac{z'_2 + vt'}{\sqrt{1-\beta^2}} \tag{2.190}$$

Simple math gives then:

$$(z'_2 - z'_1) = L\sqrt{1-\beta_{al}^2} \tag{2.191}$$

which appears shorter than the length L measured by an observer in Σ. Note that the same is true in the opposite situation, in which the observer in Σ is considered as moving relative to Σ'. Length in Σ' will appear shorter to him. If we do the calculation relative to time we find:

$$t'_2 - t'_1 = \frac{t_2 - t_1}{\sqrt{1-\beta^2}} \tag{2.192}$$

Time intervals in Σ' as observed from Σ appear to be longer than intervals measured in Σ. Seconds appear longer in Σ', as seen from Σ. However, time for an observer in Σ' appears to flow at the normal rate. It is called proper time and the observer in that system lives a normal life as if he were at rest. The observer in Σ that we claimed to be at rest makes the same observation: his time flows as usual. However, he observes that seconds in Σ' appear longer. For him, time in Σ' appears to flow at a slower rate. If the observer in Σ' returns to join the observer in Σ he will observe that his friend, left behind, has aged much faster than he did.

The postulates of relativity lead also to several important laws that we will not derive here. We will come back to the subject and derive those laws in Chapter IV by means of the principle of least action and Lagrange's equations. For example, we will find that Newton's inertial law is altered to:

$$\frac{d}{dt}\frac{m_o \dot{z}}{\sqrt{1-\left(\frac{v}{c}\right)^2}} = F(z) \tag{2.193}$$

which can be interpreted as if the mass of the object increases with its speed as observed from a system considered at rest. We thus have for a mass moving at speed v:

$$m = \frac{m_o}{\sqrt{1-\left(\frac{v}{c}\right)^2}} \tag{2.194}$$

The energy of the object in motion is readily calculated approximately as:

$$E \cong m_o c^2 + \frac{1}{2} m_o v^2 \tag{2.195}$$

where m_o is the rest mass of the object. The first term is the so-called rest mass energy of the object and the second term is the kinetic energy of the object of mass m_o moving at speed v in system Σ.

The theory of relativity makes explicit several other properties in the dynamics of the universe, often creating paradoxes. These properties are outlined in several texts and fall outside the goal of the present text. We will present some, however, in Chapter V as examples of the use of Lagrange's equations.

General Relativity

Newton's and Galileo's principles appear to describe rather well the motion of objects in our daily activities. The concept of gravity coupled with these principles provided a basis for describing mathematically the behavior of objects with mass in our terrestrial environment. Gravity and those principles were examined closely by Einstein and several others. The concept of spacetime was introduced. Einstein's approach was to consider that masses introduce a curvature of that spacetime whose main property is to provide a kind of system like a reference grid on which objects and electromagnetic waves travel. The theory is based on a system of twelve equations connecting the geometrical properties of space to its content of mass and energy. The theory leads to the same results as those obtained in Newton's approach if the motion of objects is relatively slow relative to the speed of light. However, it predicts many effects such as the deflection of light rays by massive objects, the precession of elliptical orbits, gravitational waves, the lowering of light frequency by massive objects and the expansion of the universe. This last effect is of particular interest. A solution of Einstein field equations was obtained by Friedmann for a homogeneous and isotropic model of the universe. It showed that the distance between massive objects was increasing with time as if space was expanding, as shown in Figure 2.19.

Such an expansion may be expressed as an increase in distance *l(t)* between the galaxies:

$$l(t) = a(t)l_o \tag{2.196}$$

where *a(t)* is the expansion factor. The result obtained by Friedmann can be written as:

$$\left(\frac{\dot{a}}{a}\right)^2 = \frac{8}{3}G\pi\rho + \frac{kc^2}{a^2} \tag{2.197}$$

Where the last term in the equation is a constant of integration and is interpreted as a curvature of space, ρ is the density of matter. The term on the left-hand side is the normalized expansion factor, $\frac{\dot{a}}{a}$ squared. It is called the Hubble's constant. The theory does not give its value. At present, for two galaxies separated by 1 megaparsec or 3.26 million lightyears, it is evaluated to be equal to ~72 km/s. Eq. 2.197 is an

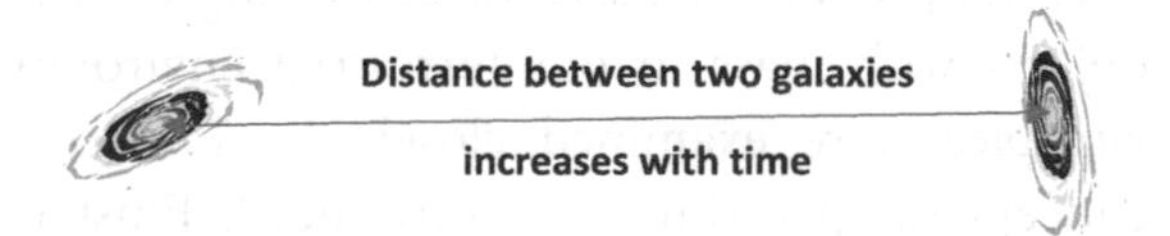
Distance between two galaxies
increases with time

FIGURE 2.19 The distance between galaxies increases as if the so-called space between them was expanding.

important result of the theory of relativity. It can also be derived from Eq. 2.95 using a simple approach based on the gravitational field only. The reader can find a demonstration in several books (see for example: Peebles, *Principles of Physical Cosmology* 1993; Vanier, *The Universe* 2011). We will show later how it can be obtained from the least action principle.

D. ELECTRO- AND MAGNETO-DYNAMICS

Maxwell's Equations

In this section we will give a short review of the main results derived in the field generally called *electromagnetism* for describing the dynamical properties of electrically charged objects, the soul of it being Maxwell's equations. We will not make mathematical demonstrations of the results obtained in that field; we will only describe those results. Demonstrations can be found in many textbooks such as those listed in the bibliography at the end of this book.

When objects contain electric charges, a force of attraction or repulsion exists between them. The force is proportional to the product of the value of the charges q and q' divided by the square of the distance r between the electrically charged objects. It is written as:

$$\boldsymbol{F} = \frac{qq'}{4\pi\varepsilon_o r^2}\mathbf{e}_r \tag{2.198}$$

and is called Coulomb's law. The force is directed along the line joining the two charges, whose direction is given by the unit vector $\mathbf{e}_r$. It is similar to the law of gravitation. However, nature has invented two types of charge which either can attract or repel each other. We have called those charges positive (+) and negative (-). Charges of opposite signs attract each other while charges of the same sign repel each other. The law above applies to charges of the same sign. The constant ε_o is called the vacuum dielectric constant and has a value of 8.85 x 10^{-12} farad/m. The factor 4π is introduced for providing simplification of notation in other equations resulting from the theory.

The size of that electric force in the case of two electrons separated, say, by 1 mm can be compared to that of gravity between the same electrons. The elementary charge of an electron called e has a value of

1.602×10^{-19} coulomb while its mass is 9.11×10^{-31} kg. Using Eq. 2.70 and 2.198, we obtain a ratio of electric force to gravitational force equal to ~ 10^{42}. For two protons in the same circumstances we obtain the ratio ~ 10^{38} since the proton has a mass larger than that of the electron by a factor of the order of 2000.

Like in the case of gravity we define an electric field E produced by a charge q' as:

$$\boldsymbol{E} = \frac{q'}{4\pi\varepsilon_o r^2}\boldsymbol{r} \tag{2.199}$$

A charge in motion produces a magnetic field $\boldsymbol{H}$. This field is also represented by the so-called magnetic induction $\boldsymbol{B}$ related to $\boldsymbol{H}$ in vacuum by the relation $\boldsymbol{B}=\mu_o\boldsymbol{H}$, where μ_o is the magnetic constant equal to 4π x10^{-7} Newton per squared meter. A charge q moving at velocity **v** and submitted to a magnetic induction $\boldsymbol{B}$ is submitted to a force given by the vector product of **v** and $\boldsymbol{B}$:

$$\boldsymbol{F} = q\mathbf{v} \times \boldsymbol{B} \tag{2.200}$$

The force is at a right angle to both vectors **v** and $\boldsymbol{B}$. For a charge q in motion submitted to both an electric field $\boldsymbol{E}$ and a magnetic induction $\boldsymbol{B}$, the total force is given by:

$$\boldsymbol{F} = q(\boldsymbol{E} + \mathbf{v} \times \boldsymbol{B}) \tag{2.201}$$

This force is called the Lorentz force and accelerates charged particles according to Newton's law. We thus have:

$$m\frac{d^2\boldsymbol{r}}{dt^2} = q(\boldsymbol{E} + \mathbf{v} \times \boldsymbol{B}) \tag{2.202}$$

giving an acceleration to the charge equal to:

$$\boldsymbol{a} = \frac{q}{m}(\boldsymbol{E} + \mathbf{v} \times \boldsymbol{B}) \tag{2.203}$$

The magnetic induction caused by a current $\boldsymbol{I}$ obeys Ampere's circuital law:

$$\oint \boldsymbol{B} \cdot \, \boldsymbol{dl} = \mu_o \boldsymbol{I} \tag{2.204}$$

where the integral is on a closed circuit enveloping the current $\boldsymbol{I}$.

It is practical to associate the magnetic field with a quantity called the magnetic flux Φ which is also a measure of its intensity. A very important phenomenon discovered by Faraday concerns the effects of fluctuations of the fields themselves with time: if the magnetic field varies with time, the flux also varies and an electric field $\boldsymbol{E}$ is induced in the surrounding space or vacuum. It is as if the vacuum, stretched by the varying flux, reacts orthogonally and creates an electric field to compensate. The size of the electric field $\boldsymbol{E}$ created is given by a formula:

$$\oint \boldsymbol{E} \cdot \boldsymbol{dl} = -\frac{d\Phi}{dt} \tag{2.205}$$

where the integral is essentially a summation of all elements $\boldsymbol{E} \bullet \boldsymbol{dl}$ over a closed circuit. These are the laws of electricity and magnetism that are at the basis of the dynamics of charged objects in the universe when electric and magnetic fields are present. On the basis of these laws, Maxwell developed an analysis that unifies completely electric and magnetic concepts into a single theory that is called "*classical electromagnetism*". The final result consists in four equations relating the vectors $\boldsymbol{E}$ and $\boldsymbol{B}$ to each other:

$$\nabla \cdot \boldsymbol{E} = \frac{\rho}{\varepsilon_o} \tag{2.206}$$

$$\nabla \times \boldsymbol{E} = -\frac{\partial \boldsymbol{B}}{\partial t} \tag{2.207}$$

$$\nabla \cdot \boldsymbol{B} = 0 \tag{2.208}$$

$$\nabla \times \boldsymbol{B} = \frac{j}{\varepsilon_o c^2} + \frac{1}{c^2}\frac{\partial \boldsymbol{E}}{\partial t} \tag{2.209}$$

The symbol ρ represents the charge density in the space where the equations are applied. The symbol $\boldsymbol{j}$ stands for the density of current due to the motion of the charges in the same space. It is possible with these equations to essentially describe all possible experimental situations providing that we know the boundary conditions. They provide a complete description of the fields $\boldsymbol{E}$ and $\boldsymbol{B}$ in the environment studied and consequently provide the basic equations for calculating the forces seen by charged particles submitted to those fields. Unfortunately, it is only possible to solve them analytically in simple symmetrical geometric situations. For more complex geometries it is necessary to use numerical solutions for which a digital computer is an ideal tool. Programs exist that solve those equations for the most complex situations, such as resonant cavities, and provide a picture of the distribution of the fields $\boldsymbol{E}$ and $\boldsymbol{B}$. This set of equations is normally completed with an equation that fulfils the condition of conservation of charges.

$$\frac{\partial}{\partial t}\rho + \boldsymbol{\nabla} \cdot j = 0 \tag{2.210}$$

This equation implies that the rate of change of the charge density ρ in one region of space is exactly equal to the current density $\boldsymbol{j}$ that leaves the region: no electric charges are created or destroyed in any region of space. Electric charge is conserved.

It may be mentioned that Maxwell's equations are not symmetrical regarding $\boldsymbol{E}$ and $\boldsymbol{B}$. In electrostatic we have the electric charge density ρ_e, but in magnetostatics we do not have the magnetic charge density ρ_m. No monopoles have yet been found in nature. We thus have the relation, $\boldsymbol{\nabla} \cdot \boldsymbol{B}$ equal to zero, which means that there are no isolated magnetic charges.

Energy is contained in electric and magnetic fields. It is given by:

$$W_{el} = \frac{e_o}{2} \int_{\mathrm{V}} E^2 dv \tag{2.211}$$

$$W_{mag} = \frac{1}{2\mu_o} \int_{\mathrm{V}} B^2 dv \tag{2.212}$$

The integrals are done over the volume desired, delimitating the region of interest.

Electromagnetic Wave Equation

A combination of Maxwell's equations and some algebra leads directly to the wave equation

$$\left(\frac{\partial^2}{\partial x^2}+\frac{\partial^2}{\partial y^2}+\frac{\partial^2}{\partial z^2}\right)\boldsymbol{E}=\frac{1}{c^2}\frac{\partial^2}{\partial t^2}\boldsymbol{E} \tag{2.213}$$

with c, the speed of light, given by

$$c^2=\frac{1}{\varepsilon_o\mu_o} \tag{2.214}$$

A similar expression is obtained for the magnetic field $\boldsymbol{H}$. A solution of Eq. 2.213 leads to waves propagating in free space at speed c, as given by the expressions:

$$\boldsymbol{E}=\mathbf{i}E_o cos(\omega_k t-\boldsymbol{k}\cdot\boldsymbol{r}) \tag{2.215}$$

$$\boldsymbol{H}=\mathbf{j}H_o cos(\omega_k t-\boldsymbol{k}\cdot\boldsymbol{r}) \tag{2.216}$$

where **i** and **j**, unit vectors at a right angle to each other, make the magnetic field $\boldsymbol{H}$ perpendicular to the electric field $\boldsymbol{E}$. ($\omega_\kappa = 2\pi\nu_\kappa$) is the angular frequency of the wave identified by the propagation vector $\boldsymbol{k}$ along direction $\boldsymbol{r}$ and equal to $2\pi/\lambda$. We also have the relation $\nu\lambda=c$. The wave described by Eqs. 2.215 and 2.216 is said to be linearly polarized, the electric field $\boldsymbol{E}$ being along the unit vector **i**, the x direction, while the magnetic field $\boldsymbol{H}$ is at a right angle along the unit vector **j** and oscillates in phase with it. In the case of a circularly polarized wave, there is a $\pi/2$ phase shift between $\boldsymbol{E}$ and $\boldsymbol{H}$ and the electric field $\boldsymbol{E}$ appears to rotate around the direction of propagation.

We note that the speed of light c appears naturally in Maxwell's equations as a combination of the two constants ε_o and μ_o characterizing a vacuum, presumably a "non-substance". That speed is independent of the reference frame that may be selected. It has a nature very different to that of the speed of sound in air, which reflects the physical characteristics of air. If air moves, sound waves, which are pressure fluctuations, follow it and their speed relative to a fixed reference frame is altered. Electromagnetic radiation propagates in a vacuum at speed c independently of the frame of reference in which that speed is measured.

E. QUANTUM PHYSICS AND ELECTRODYNAMICS

Schrödinger's Equation

Quantum mechanics started with the assumption that the variation in the intensity of radiation emitted by a so-called blackbody is a discontinuous process. Before the introduction of that concept it was not possible to explain the shape of the spectrum of radiation emitted by a body at a given temperature. Planck solved the problem by assuming that variation of the intensity of electromagnetic radiation was not continuous. Energy was emitted as discrete quantities having small but finite dimensions. This is one of the most important steps in physics in the understanding of the dynamics of radiation emission by objects at any temperature above 0 K. For the first time, the concept of discontinuity or quantization in the variation of the intensity of electromagnetic radiation was introduced. That concept was used for introducing the concept of the photon of energy $h\nu$ or hc/λ and it opened the door to the introduction of the field of quantum mechanics.

Quantum mechanics, one of the strangest theories ever developed, has provided us with a rather good model of the nature of atomic particles and the propagation of electromagnetic radiation. Our understanding of the dynamics of the electron originates from the proposal that it can be represented either by a wave or a particle, depending on the circumstances. This is a postulate that we have to accept. Actually, we do not understand the matter inner gears that give that property to all particles. As a wave, its behavior is described by means of the so-called wave equation known also as Schrodinger's equation, written as:

$$\left(-\frac{\hbar^2}{2m}\nabla^2 + V\right)\psi = i\hbar\frac{\partial}{\partial t}\psi \tag{2.217}$$

where m is the mass of the electron and ψ is the so-called wave function that describes the behavior of the electron in the potential V. This equation reflects effectively the difference between the quantum world and the classical fields derived from Maxwell's wave equations and given by Eqs. 2.206 to 2.209. When the equation is solved for the case of the hydrogen atom, the electron behavior is described successfully, its energy being quantized and limited to specific values.

That equation may be accepted as such, somewhat like Newton's law describing the effect of a force on a mass m in a gravitational field. Nevertheless, that equation can be obtained directly from operator algebra, a technique used at large in quantum mechanics. The idea is to interpret some classical observable as operators and give them a form according to specific rules. This is achieved in the following way. Earlier we introduced the total energy E of a particle as the sum of its kinetic energy T and its potential energy V, written as:

$$\frac{p^2}{2m} + V = E \tag{2.218}$$

This equation can be used as an operator applied to the wave function ψ of the particle. The question is then: what is the form of those operators representing the classical variables p, V and E of that equation? They are:

$$p_x = -i\hbar\frac{\partial}{\partial x} \tag{2.219}$$

with similar expressions for p_y and p_z. For E we adopt as operator:

$$E = i\hbar\frac{\partial}{\partial t} \tag{2.220}$$

and V stays as V. When these operators are replaced in Eq. 2.218 operating on ψ, one obtains Eq. 2.217. When we say that quantum mechanics is a strange theory, this derivation and Eq. 2.217 that dictates the behavior of an atomic particle through the wave function ψ and complex (imaginary) operators is a rather convincing demonstration. The main argument that confirms its validity is that it agrees with experimental observations.

In the early days after the development of that theory, questions were raised relative to the physical meaning of ψ, the wave function called also material wave. It has several specific properties. A proposal made by Max Born in an effort to give a physical meaning to that wave function has survived to today. It is assumed that its squared absolute value, $|\psi|^2$, multiplied by an element of volume "dv", represents the probability of finding the electron in that given element of volume. It is thus essentially a mathematical tool used to represent the probability of finding the position of the electron. Consequently, the electron is not localized precisely. This uncertainty is best expressed in the uncertainty relation proposed by the Heisenberg principle $\Delta x \Delta p > \hbar$, which says that position and momentum can be determined only within a limited precision. Similarly, the concept applies between time t and energy E, expressed as $\Delta E \Delta t > \hbar$. The energy of an elementary object that has a lifetime Δt cannot be determined to better than $\hbar/\Delta t$. On the other hand, that property of the quantum world gives rise to the possibility of the existence of virtual particles if their lifetime is very short. In a sense energy is borrowed from a vacuum to create particles, with $m = E/c^2$, and is reimbursed within a very short period. It is like borrowing money from the bank for a time so short that the accounting system does not have time to detect it. If the sum is reimbursed quickly there are no consequences.

The Photon Concept

Electromagnetic waves were described in a classical sense above. When such a wave is maintained in an enclosure of volume V, the total energy of the electromagnetic field in the volume is given by:

$$Total\ energy = \frac{1}{2}\int_V \left(\varepsilon_o \boldsymbol{E}\cdot\boldsymbol{E} + \frac{\boldsymbol{B}\cdot\boldsymbol{B}}{\mu_o}\right)dv \qquad (2.221)$$

The fields $\boldsymbol{E}$ and induction $\boldsymbol{B}$ can be expressed as a summation over modes within the volume V, which can be considered as a cavity. When this is done an equation is obtained that assimilates the behavior of the fields to that of a harmonic oscillator. The fields are then represented by means of normal modes and creation and annihilation operators (see, for example, *The Quantum Physics of Atomic Frequency Standards*, Vanier and Audoin, p. 121). The energy within those fields is then given by:

$$Energy\ in\ the\ wave = \left(n + \frac{1}{2}\right)h\nu \qquad (2.222)$$

where n is interpreted as the number of particles of energy hv in the wave. Those particles are called photons. The fraction ½ that appears in the equation is interpreted as zero-point energy. It is energy attached to a physical system and cannot be extracted. It is most interesting that those photons keep all the properties of the supporting electromagnetic wave, such as frequency and phase. The wave is then said to be made of n photons, that number being a measure of the intensity of the excitation of the electromagnetic wave. Photons going through slits keep phase information and produce interferences.

Quantum Mechanics and Relativity

When special relativity is introduced in Eq. 2.217 as was achieved by Dirac, it is found that the electron possesses an intrinsic angular momentum called "spin" with its associated magnetic moment. The analysis incorporating relativity also shows the possibility of the existence of particles identical to the electron but with a positive charge. They look like electrons running backward in time. They were given the name positrons. The existence of such a particle was demonstrated in 1932 by Anderson. It is an antiparticle leading to the concept of antimatter. It should also be mentioned that a vacuum has the property of generating pairs of elementary particles for very short times, called virtual particles. Those particles are not detected as such but their presence, for example, is confirmed by their effects on observed optical spectra.

We will not expand further on the subject of quantum physics. The short elementary introduction just given is sufficient for the present purpose, which is to outline the main physics based on Schrödinger's equation. We will show in Chapter IV that we can obtain the same result using Lagrange's approach.

CHAPTER III

Search for a Universal Principle

In the previous chapter, we conducted an elementary analysis of the behavior of objects with mass and electric charge in the presence of gravitational and electromagnetic fields. We analyzed the dynamics of those objects submitted to the forces generated by those fields. The analysis was based on Newton's laws and Maxwell's equations, and the concepts of forces and action that cause reaction. We conducted several analyses using these laws in specific situations, and we saw that we could describe the behavior of masses and electrically charged objects with those laws rather well. We also introduced some other concepts such as the conservation of certain mechanical properties like energy, linear and angular momentum, as well as the conservation of electrical charges. We have done that in a 3-D space environment and use vector algebra.

We also introduced the concept of quantization, which appears to be necessary to explain the behavior of objects and fields at the atomic level. In doing so, we saw that electromagnetic fields have a quantum character that makes them act as propagating particles called photons, and we introduced the concept that particles have a propagating property that makes them act as a wave. The dynamics of atomic particles are then represented by a wave equation that can be obtained in an abstract way by replacing classical operators in the equation representing the total energy of the particle by quantum operators. That equation, popularly known as Schrödinger's equation, provides a basic tool by means of which the dynamics of the atomic world can be analyzed. All those laws seem to be independent of each other. We also have introduced conservation laws that appear to be conceptual, but which can be used in deriving rules that objects obey in their evolution. These rules certainly seem to work rather well when explaining the behavior of objects and fields in the universe.

We are allowed to question the universality of those laws. It is legitimate to ask ourselves if the laws, as introduced by Newton, Maxwell and Schrödinger, are as basic as we can intuitively conclude. Furthermore, those laws appear to be quite arbitrary and we may ask ourselves if there is not a fundamental principle that could be used to arrive at the same results regarding the motion of objects in space and conservation laws, or even provide a basis for the quantum behavior of the universe at the small-scale level.

It is a question that has intrigued human beings since they first started reflecting on the exact nature of that physical universe we inhabit. No general answers to that profound question covering the entire field of physics have been found. There are many aspects of the behavior of objects in the universe that lead us to believe that the universe may evolve by means of a basic principle that is more general than the specific laws enunciated above. For example, it looks very much that in general nature favors situations of minimum energy. Objects falls on the ground where energy is lower. In searching for minimum energy, nature appears also to look for equilibrium or stationarity. Another observation is that the universe appears to behave in a manner that tends towards simplicity. For example, paths of objects thrown in the air above the earth's surface seem to be rather regular and uniform. Light travels in a straight line and not in a complicated path; it is reflected from surfaces with a behavior that has been put under the form of rather simple rules. The conservation principles introduced earlier appear to have some origin that deserve deeper reflection than just being the result of assumptions and the elementary analysis that we made. All these considerations lead us to believe that nature is driven by some basic law or principle that just needs to be discovered.

Such a principle was proposed some time ago following the deep thinking of physicists that followed Newton, such as Bernoulli, Euler, Lagrange, Hamilton and several others. The principle in question is called *principle of least action.* We will see that a study of that principle leads to differential equations called *Lagrange's equations*, which essentially replaces the dynamical equations of motion and conservation laws that we have introduced above. When that principle is applied properly, it also leads to basic laws governing the evolution of the universe as regards its general behavior as well as some of its details. In fact, the approach using that principle is rather universal and appears to regulate the dynamics of *material objects* and entities such as *fields* within the whole of physics at both the large scale as well as at the atomic level.

Before going into the mathematical description of that principle, we need, however, to familiarize ourselves with an interesting mathematical tool called the *calculus of variations*. Once we understand that mathematical tool we will be in a better state to understand the meaning of that principle and related calculations.

A. A MATHEMATICAL TOOL: CALCULUS OF VARIATIONS

In order to understand a particular mathematical tool, it is best to perform the analysis directly with a concrete example. Following that approach, let us attempt to determine the path to be followed by an object in space between two fixed points, with the requirement that that path be the *shortest distance* between the two points. This is an interesting problem. For example, such a path is followed by an airplane travelling across an ocean between two continents in order to minimize time of flight and fuel consumption. The airplane travels on the surface of a sphere. That surface envelopes the earth and is closed in on itself. It is not evident which of the various possible paths on that sphere is the shortest one. It is not the path forming a straight line when we look at the surface of the globe as represented on a plane as in a Mercator projection. We know by experience that it is the path that follows a great circle around the globe. But to prove it mathematically is a different matter.

To familiarize ourselves with the analysis, we look first at the same exercise on a Cartesian flat plane as shown in Figure 3.1. To determine the shortest path between points A and B, we need an equation that describes a general path between the two points and then we must analyze that equation. In Figure 3.1, three paths called C_1, C_2 and C_3 passing through those points are shown.

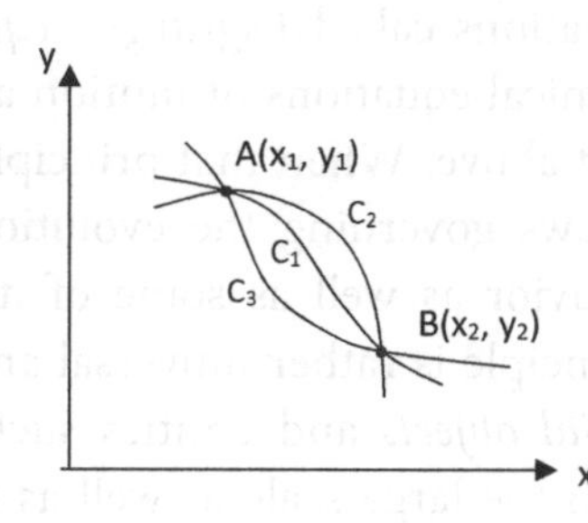

FIGURE 3.1 Illustration showing three paths that join points A and B.

The curves are forced to pass through points A and B at coordinates (x_1, y_1) and (x_2, y_2).

Curve C_2 could be a parabola for which we can write a simple second-order equation in terms of coordinates x and y. Like the two other paths it has nothing special. We can draw several other paths passing through points A and B and we could in principle represent them also by specific equations. The length of each of those paths can be calculated by integrating each equation over their respective path from A to B. We would then obtain a number for each path, which would enable us to make a table using those numbers and decide, by examination of the table, which one is the shortest path between A and B. This would be a lengthy procedure. We know of course that, in that simple case, the path of minimum length is a straight line between the two points. The plane on which the trajectories are drawn is flat and, in practice, intuition is sufficient to decide which is the shortest path. If the plane is not flat, such as the surface of a globe, the solution is not obvious. Is there a general rule by which we can find a curve $y(x)$ that minimizes the length of the path? The answer is yes and we can formulate that rule mathematically. The technique is called "*calculus of variations*". To do the analysis, we use a flat plane as illustrated in Figure 3.2 and assume that a function f exists that represents curves that can be written to represent paths from A to B. The function f is unknown.

While one of the curves could be represented by an equation such as $y(x)$ and is a function of x, f is a function of the curves $y(x)$ and can be written as $f(y(x))$. Thus, the length of the path is determined by an integral over that undetermined function f over the path from A to B. We call the length of that path S, which is given by:

$$Length\ of\ path = S = \int_A^B f(y(x))dx \tag{3.1}$$

This integral gives a number for every *y(x)* chosen and *S* is a function of the function *f(y(x))*. It is called a functional. What we need to do is to minimize that *path length, S*, that varies with the *y(x)* chosen. From our experience with differential calculus, we know that in order to find the extremum (maximum or minimum) of a function representing a curve in space, we only need to differentiate that function, equate the result to zero and solve the resulting equation. However, here, it is not the function itself that must be extremized. It is the length of the path as given by Eq. 3.1 that needs to be extremized. The function *y(x)* has a numerical value at each point, while the total path is an integral over the infinitesimal elements of that path. Let us make a detailed calculation of that path length.

As shown in Figure 3.2, a length element *dS*, along the arbitrary curve shown, is given by:

$$dS^2 = dx^2 + dy^2 \quad \text{or} \quad dS = \sqrt{dx^2 + dy^2} \tag{3.2}$$

Our integral over the path can thus be written:

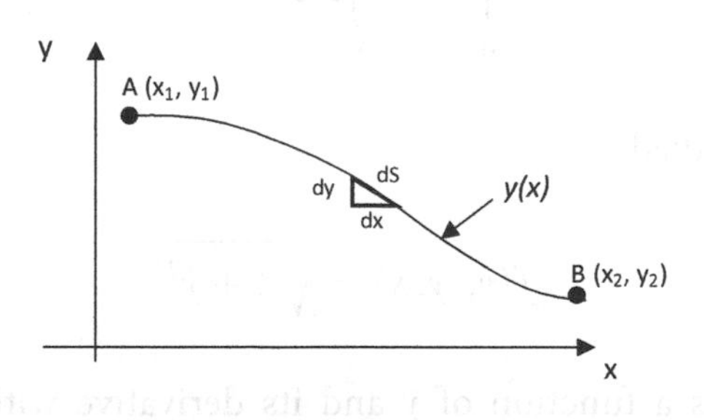

FIGURE 3.2 Diagram representing a possible path *y(x)* for travelling from point A to point B in a 2-D flat space.

$$S = \int_A^B dS = \int_A^B \sqrt{dx^2 + dy^2} \tag{3.3}$$

On the other hand, using x as a parameter, we can write Eq. 3.2 as:

$$dS = \sqrt{1 + \frac{dy^2}{dx^2}}\, dx \quad \text{or} \quad dS = \sqrt{1 + \dot{y}^2}\, dx \tag{3.4}$$

where, in order to simplify notation, we have used the derivative notation:

$$\dot{y} = \frac{dy}{dx} \tag{3.5}$$

This notation should not be confused with a differentiation with respect to time, which will be used later on. The integral thus becomes:

$$S = \int_{x_1}^{x_2} dS = \int_{x_1}^{x_2} \sqrt{1 + \dot{y}^2}\, dx \tag{3.6}$$

with x as variable of integration. We can thus write this as:

$$S = \int_{x_1}^{x_2} dS = \int_{x_1}^{x_2} f(y,\ \dot{y}, x)\, dx \tag{3.7}$$

where f is identified as

$$f(y,\ \dot{y}, x) = \sqrt{1 + \dot{y}^2} \tag{3.8}$$

We see that f is a function of y and its derivative with respect to the parameter x. This is an important point to remember. If we want to find the extremum of S, as represented by such an expression, we need to differentiate it and, for this, we need to know *f(y(x))*. We only know it in terms of the derivative of *y* with respect to *x*. However, if *S* is an extremum, we know that, at that extremum, a small change in the coordinates will result in a zero first order change in the function. To do this we may alter *y* by a small quantity, say $\varepsilon\eta(x)$, and calculate its effect on *S*. We write this change as:

$$y(x,\ \varepsilon) = y(x,\ 0) + \varepsilon\eta(x) \tag{3.9}$$

where $\eta(x)$ is an arbitrary function that vanishes at the two points A and B, since we want the path to pass through those points. Thus, the integral, Eq. 3.7, is function of the small change ε that we introduce in the coordinate *y*.

$$S(\varepsilon) = \int_{x_1}^{x_2} f(y(x,\varepsilon), \dot{y}(x,\varepsilon), x)\ dx \tag{3.10}$$

We want to know the rules that *f* has to obey to make *S(ε)* extremum. We thus assume that *S(ε)* varies continuously with ε. It is a function that possibly shows an extremum (either a minimum, a maximum or a saddle point). To find that extremum we have to differentiate *S(ε)* relative to ε, and then equate the result to 0. We can thus write:

$$\frac{dS(\varepsilon)}{d\varepsilon} = \frac{d}{d\varepsilon}\int_{x_1}^{x_2} f(y(x,\varepsilon), \dot{y}(x,\varepsilon), x)dx = 0 \tag{3.11}$$

We can differentiate under the integral sign using standard rules of partial differentiation with various independent variables. We do it in the following manner:

$$\frac{\partial S(\varepsilon)}{\partial \varepsilon} = \int_{x_1}^{x_2} \left(\frac{\partial f}{\partial y}\frac{\partial y}{\partial \varepsilon} + \frac{\partial f}{\partial \dot{y}}\frac{\partial \dot{y}}{\partial \varepsilon} \right) dx \tag{3.12}$$

We have kept x constant independent of ε as mentioned above, changing only y by the small quantity $\varepsilon\eta$. The second term under the integral sign, using Eq. 3.12, may be written as:

$$\int_{x_1}^{x_2} \frac{\partial f}{\partial \dot{y}} \frac{\partial \dot{y}}{\partial \varepsilon} dx = \int_{x_1}^{x_2} \frac{\partial f}{\partial \dot{y}} \frac{\partial^2 y}{\partial x \partial \varepsilon} dx \tag{3.13}$$

We integrate this second term by means of the rules of integration by part and we obtain:

$$\int_{x_1}^{x_2} \frac{\partial f}{\partial \dot{y}} \frac{\partial^2 y}{\partial x \partial \varepsilon} dx = \left. \frac{\partial f}{\partial \dot{y}} \frac{\partial y}{\partial \varepsilon} \right|_{x_1}^{x_2} - \int_{x_1}^{x_2} \frac{d}{dx}\left(\frac{\partial f}{\partial \dot{y}}\right) \frac{\partial y}{\partial \varepsilon} dx \tag{3.14}$$

However, a condition imposed on our calculation is that all possible curves must pass through the two points (x_1, y_1) and (x_2, y_2). Consequently, the first term on the right-hand side of the integral, evaluated at x_1 and x_2, is zero since we require that y does not vary with ε at those points and $\left(\frac{\partial y}{\partial \varepsilon}\right)_{x_{1,2}} = \eta(x_{1,2}) = 0$. Eq. 3.12 becomes:

$$\frac{\partial S(\varepsilon)}{\partial \varepsilon} = \int_{x_1}^{x_2} \left(\frac{\partial f}{\partial y} - \frac{d}{dx} \frac{\partial f}{\partial \dot{y}} \frac{\partial y}{\partial \varepsilon}\right) dx \tag{3.15}$$

On the other hand, a small change, such as δS in S and δy in y, as a function of ε, evaluated for $\varepsilon = 0$, can be written as:

$$\delta S = \left(\frac{\partial S}{\partial \varepsilon}\right)_0 d\varepsilon \qquad \delta y = \left(\frac{\partial y}{\partial \varepsilon}\right)_0 d\varepsilon \tag{3.16}$$

We multiply Eq. 3.15 by $d\varepsilon$ and use these definitions. Since S must be an extremum, δS must not change in first order with ε at $\varepsilon = 0$ and consequently $\delta S = 0$.

$$\delta S = \int_{x_1}^{x_2} \left(\frac{\partial f}{\partial y} - \frac{d}{dx} \frac{\partial f}{\partial \dot{y}} \right) \delta y dx = 0 \tag{3.17}$$

However, δy is arbitrary and thus to satisfy that condition, the term in parentheses within the integral must be 0. We can thus write:

$$\frac{\partial f}{\partial y} - \frac{d}{dx} \frac{\partial f}{\partial \dot{y}} = 0 \tag{3.18}$$

This is a differential equation providing a means of finding the function f that makes the distance between the two points (x_1, y_1) and (x_2, y_2) extremum. We started by assuming that such a function exists. We did not know what it was, but we forced it to be such in order to make extremum the integral of the path between the two points A and B in Figure 3.2. We do not know yet if that extremum is a maximum or a minimum. This final conclusion can be reached by analyzing the context of the situation in question. Thus Eq. 3.18 sets the condition that f must obey for the path between A and B to be an extremum in the present case a minimum. The calculation does not determine f, but results in an equation setting a condition on f. Eq. 3.18 is a most fascinating result. In order to familiarize ourselves with it, let us examine it using a few specific cases.

Shortest Path on a Plane

As an example, let us calculate the actual minimum path between points A and B in the plane shown in Figure 3.2. We have Eq. 3.18 and the definition of f given by Eq. 3.8. We thus calculate the required derivatives as:

$$\frac{\partial f}{\partial y} = 0 \quad and \quad \frac{\partial f}{\partial \dot{y}} = \frac{\dot{y}}{\sqrt{1 + \dot{y}^2}} \tag{3.19}$$

Eq. 3.18 reduces then to one term:

$$\frac{d}{dx}\frac{\partial f}{\partial \dot{y}} = \frac{d}{dx}\frac{\dot{y}}{\sqrt{1+\dot{y}^2}} = 0 \tag{3.20}$$

Consequently, the term under the differential relative to x must be a constant for this equation to be valid and we write:

$$\frac{\dot{y}}{\sqrt{1+\dot{y}^2}} = constant \tag{3.21}$$

This means that $\dot{y}$ must also be a constant that we make equal to a. Thus, we have:

$$\dot{y} = \frac{dy}{dx} = a \text{ or } dy = adx \tag{3.22}$$

Integrating this equation, we obtain:

$$y = ax + b \tag{3.23}$$

a straight line, as we expected. a and b are determined by means of initial conditions, that is the position of points A and B in space. As we can see, other curves do not satisfy the extremum condition. In the present case, it obviously shows that the extremum is a minimum. It may be concluded that it seems to be a fair amount of work for obtaining something obvious. However, many other, more complicated problems can be solved directly using this technique. Let us analyze, for example, the case of the geodesic or the shortest path between two points A and B on a sphere.

Shortest Path on a Sphere

The problem is represented in Figure 3.3. In view of the symmetry we use polar coordinates.

We want to find the shortest path between points A and B on the sphere of radius R, at respective coordinates (θ_1, φ_1) and (θ_2, φ_2). This is called a *geodesic* and is the path generally followed by airplanes to save time and fuel. We use a similar approach as in the case of the path

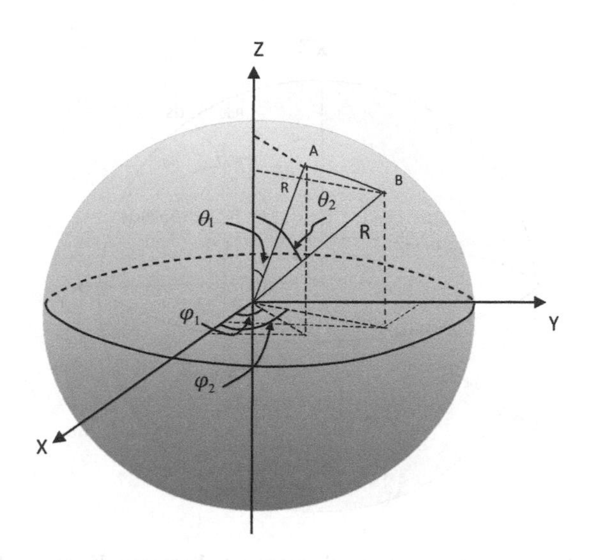

FIGURE 3.3 Representation of a path A–B in polar coordinates on the sphere of radius *R*.

on a plane. Referring to Figure 3.4, we note that the radius R is fixed and an element *dS* on path A–B is given by:

$$dS = \sqrt{R^2 d\theta^2 + R^2 sin^2\theta d\varphi^2} \tag{3.24}$$

The length of the path between A and B is given by:

$$I = \int_A^B dS = R \int_A^B \sqrt{d\theta^2 + sin^2\theta d\varphi^2} \tag{3.25}$$

or still:

$$I = \int_A^B dS = R \int_A^B \sqrt{1 + sin^2\,\theta\,\dot{\varphi}^2}\ d\theta \tag{3.26}$$

where we have used θ as a parameter and have defined:

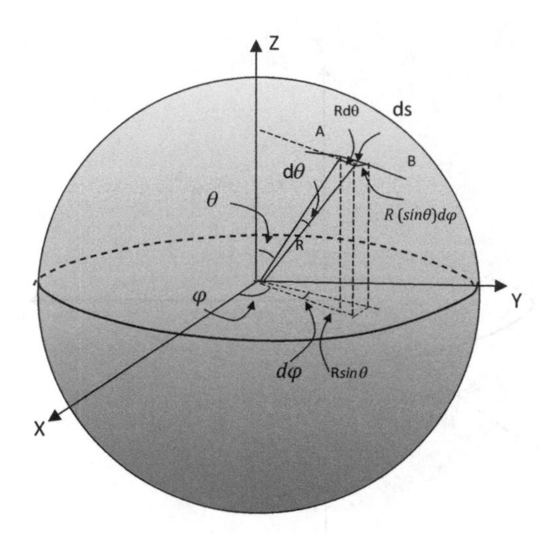

FIGURE 3.4 Representation of an element of distance dS on path A–B, on a sphere of radius R.

$$\dot{\varphi} = \frac{d\varphi}{d\theta} \tag{3.27}$$

We thus have an equation similar to Eq. 3.7 but with polar coordinates θ, ϕ and $\dot{\varphi}$:

$$I = \int_A^B f(\phi, \dot{\varphi}, \theta)\, d\theta \tag{3.28}$$

and with $f(\phi, \dot{\phi}, \theta)$ given by:

$$f(\varphi,\ \dot{\varphi},\ \theta) = R\sqrt{1 + sin^2\ \theta\ \dot{\varphi}^2} \tag{3.29}$$

In order for the path to be an extremum (minimum or maximum), we only have to require that $f(\phi, \dot{\phi},\ \theta)$ satisfy Eq. 3.18 for polar coordinate ϕ, with θ as a parameter or

$$\frac{\partial f}{\partial \varphi} - \frac{d}{d\theta}\frac{\partial f}{\partial \dot{\varphi}} = 0 \tag{3.30}$$

It is noted immediately that f is independent of ϕ. Consequently, we are left with only the second term, which is equal to 0. For this condition to be valid we must have $\frac{\partial f}{\partial \dot{\varphi}}$ equal to a constant. Thus, using Eq. 3.29, we have:

$$\frac{\partial f}{\partial \dot{\varphi}} = R\frac{sin^2\theta\dot{\varphi}}{\sqrt{(1 + sin^2\theta\dot{\varphi}^2}} = constant = c \tag{3.31}$$

This is all that is obtained from the analysis. There are many avenues that can be used to obtain a conclusion on the actual path. The algebra is a little involved but straightforward. Let us try to do it as simply as possible. First, we square Eq. 3.31 and with some algebra we get:

$$\dot{\varphi} = \frac{d\varphi}{d\theta} = \frac{a}{sin\theta(sin^2\theta - a^2)^{\frac{1}{2}}} \tag{3.32}$$

where a stands for c/R. This equation can be integrated using trigonometric relations between $sin\theta$, $cot\theta$, $csc\theta$ and their derivatives. One obtains:

$$\varphi = cos^{-1}\frac{\text{a cot}\theta}{\sqrt{(1 - a^2}} + c_2 \tag{3.33}$$

where c_2 is a constant of integration. Further algebraic manipulations give finally:

$$R\ cos\theta = AR\ sin\theta\ cos\varphi + BR\ sin\theta\ sin\varphi \tag{3.34}$$

where A and B stands for:

$$A = \frac{(1 - a^2)^{\frac{1}{2}}}{a}\cos c_2 \qquad B = \frac{(1 - a^2)^{\frac{1}{2}}}{a}\sin c_2 \tag{3.35}$$

From Figure 3.4 we recall that:

$$z = R\ cos\ \theta,\ X = R \sin\theta \cos\varphi\ and\ y = R \sin\theta \sin\varphi \tag{3.36}$$

with the final result:

$$z = Ax + By \tag{3.37}$$

This is the equation of a plane passing through the origin $z= x= y = 0$ and intersecting the sphere at radius R. Consequently, the path is on both this plane and the sphere, as was expected. The intersection of this plane with the sphere describes a great circle or a *geodesic*, which dictates the shortest path between two points situated on that intersection.

We can actually check directly on the coherence of the approach by calculating the actual length of a specific path between two angles θ_1 and θ_2. We replace ϕ in Eq. 3.26, by its value in Eq. 3.32, use Eq. 3.35 and perform the integration. We obtain:

$$I = R\left(cos^{-1}\left(\frac{\cos\theta_2}{(1-a^2)^{\frac{1}{2}}}\right) - cos^{-1}\left(\frac{\cos\theta_1}{(1-a^2)^{\frac{1}{2}}}\right)\right) \tag{3.38}$$

The constant "*a*" needs to be determined. Let us look at a specific case. Say that point A is on the z axis and point B is in the xy plane. The integral is independent of ϕ and we have $\theta_1 = 0$ and $\theta_2 = \pi/2$. In that case, we know the distance between the two points to be:

$$I = R\ \frac{\pi}{2} \tag{3.39}$$

That is the result expected. The constants *a* and *c* are given by

$$c = aR = R\sqrt{\left(1 - \frac{4}{\pi^2}\right)} \tag{3.40}$$

which determines the value of *c*, the first constant introduced by means of Eq. 3.31. This example, although somewhat involved in terms of trigonometric relations, shows the power of the calculus of variations tool.

B. THE PRINCIPLE OF LEAST ACTION AND LAGRANGE'S EQUATIONS

The search for a new principle of physics requires a lot of imagination. For example, just writing Newton's second principle that says that acceleration is force divided by mass was a rather audacious step at the time Newton introduced that second law. First, it requires the introduction of new concepts such as mass and force. Second, very often it requires the rejection of concepts accepted as evidence during previous centuries. For a long period, it was assumed that an object in movement requires the application of a force to keep it in motion at constant speed. The presence of friction in most experiments certainly helped in justifying that conclusion. If friction is present and not recognized, wrong conclusions are easily drawn. Newton's principles assume that an object in movement in a given coordinate system with an absence of friction is a natural state, one that does not require a force to maintain it in that state. It showed that an applied force changes its velocity. Furthermore, uniform motion is relative, as explained above; the state of rest is arbitrary and depends on the state of motion of the observer.

On the other hand, we can ask what path the universe chooses in the evolution of an object or system from one state to another. To simplify the question, we can limit our analysis to the movement of a particular object or system in its own configuration space in isolation from the rest of the universe. In its motion from point A to point B, or from time t_1 to time t_2, an object follows a given path depending on the forces applied to it and the initial conditions. What determines that path? We could calculate it from Newton's principles as we did earlier. In one example, we did it for orbits in a gravitational field. However, is there not something more fundamental that dictates the path to be followed by objects submitted to a force such as gravity, for example? This is a legitimate question. Does the universe take paths that optimize trajectories in space or in time in order to minimize or extremize some internal properties? We analyzed in the previous section the question of the shortest path between two points in space by means of the so-called calculus of variations. We found that the function describing that actual shortest path must obey special conditions given by Eq. 3.18. Does nature wish to do a similar thing, that is to say, take the shortest, easiest or most efficient path between two situations or states? This is not at all obvious. The only observation that we can make at this point is that the universe follows Newton's laws in its evolution and that, whatever new principle we may introduce, the results deduced from its application to specific situations must agree with those obtained when Newton's laws are used.

We realize first that the universe seems to evolve in such a way as to maintain an equilibrium between the forces applied and the effect resulting from those forces. For example, if a particle is submitted to the force of gravity it is accelerated and it follows a determined path. At every instant during that acceleration the term "*ma*" may be considered an inertial force that opposes and is equal to the force applied, "*F*". This is a conclusion drawn from Newton's second law. Consequently, during the whole path there is an equilibrium between the two forces and there is a response from the part of the particle that takes a definite path to obey that law. If we repeat that experiment, the particle follows exactly the same path. There is something special about that path that makes it immutable. It looks as if the path is optimized. If so, we could in principle determine the path by means of a calculus of extrema. We can then say that the path appears to be stationary. This can be interpreted in a very broad sense as if nature is constantly maintaining equilibrium at each instant of the path followed by the particle and that it is sniffing its way through space. We can expand that concept to the evolution of the whole universe. If we accept that kind of reasoning, we can certainly make a principle of it. But what can we optimize in that path? For the purpose of answering that question the concept of "*action*" was introduced. We have encountered that concept in the case of Newton's third law. A force causes an action and there is a reaction to that force. In common language, as its name indicates, action appears to characterize some kind of integral property of the motion of the object. We may define that action in configuration space as an intrinsic characteristic of a physical evolution, for example, of the total path followed by an object or system, between two states or two instants of time. But how do we define it mathematically? Since action characterizes the whole path between two states, we define it through an integral of a chosen parameter over the whole path, in a similar way as in Eq. 3.7 for the case of shortest path on a plane or a curved surface. In that case, it was the total length of the path that was calculated by means of integration and the actual function f inside the integral needed to be selected or calculated, thus satisfying Eq. 3.11. On its side, nature may wish to do something similar, for example use the easiest or shortest path, whatever that means. From the experience obtained above by means of the calculus of variation, the obvious thing to do is to extremize that action. Whatever the function chosen, we have to make sure it gives results compatible with Newton's principles. Thus, the main problem is to select the appropriate parameter to define the action.

In early studies on this matter, momentum, speed and some other parameters were selected for that function, but were later abandoned. We introduced earlier the concept of total energy as the sum of kinetic energy and potential energy: $E = T + V$. That quantity is conserved and is most interesting in setting initial and final conditions in a particular process. It is extensively used in thermodynamics, which studies exchanges of energy of various forms. Having a constant value in a given process, total energy does not appear, however, to be useful here for optimizing a path. On the other hand, in our experiment with an object thrown up in the air we saw that there was a continuous exchange between kinetic and potential energy, even though the total energy is conserved. That difference between these two energies has been called the *Lagrangian*, denoted L. In mechanics, it is written as:

$$L = T - V \tag{3.41}$$

Could that term be used for defining the action? At first sight, such a concept appears to be rather imaginative and does not look very promising for characterizing the functioning of the universe. We already know the behavior of T in response to V from Newton's laws. Would extremizing the action defined with this function as argument give the same results as Newton's laws? This operation means defining action, which we called J, as:

$$action = J = \int_{t_1}^{t_2} L dt \tag{3.42}$$

In that context, action has the dimension: energy x time [joule x second]. This is, indeed, a rather peculiar definition with strange units.

A representation of the evolution of an object between times t_1 and t_2 in configuration space is shown in Figure 3.5.

If we accept that definition for the action, we need to extremize that expression, which means that we must request that action, when optimized, to be invariant in first order under a small change in the coordinates in which L is expressed. We can write this condition simply as:

$$\delta J = \delta \int_{t_1}^{t_2} L dt = 0 \tag{3.43}$$

That equation just says that nature chooses a path, which, when there is a small change in the parameters defining L, will cause no change in the total action when evaluated over the total path covered by the integral. It is thus essentially a stationary situation, either a maximum, a minimum or a saddle point. It is known as the *principle of least action*. It is also known as Hamilton's principle. Least action may be a misnomer, however, since the differentiation does not necessarily lead to a minimum. We need to keep that in mind although in most cases studied the action, if it obeys the condition indicated by Eq. 3.43, will be minimum.

What are the consequences of the condition imposed on J by Eq. 3.43? We can do exactly as we did in the illustration we used in

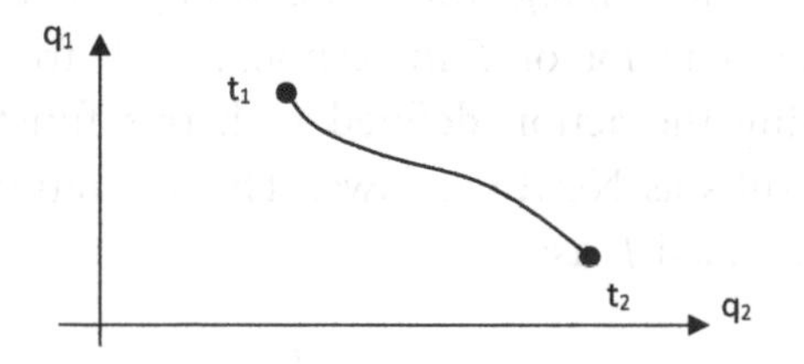

FIGURE 3.5 One of the paths that can be followed by the system between times t_1 and t_2, represented here in a 2-D configuration space.

the introduction of the calculus of variations. We introduced a small change ε in the coordinates of the function f and assumed that the differential of the integral introduced by that change would be zero at the extremum. It is important to point out that the specific definition of L in terms of kinetic and potential energy as given by Eq. 3.41 dos not enter in the calculation. This is most interesting. We will see below, however, how that question is addressed.

At this stage, we pass to a system of generalized coordinates. In fact, L can be expressed in terms of several different coordinates depending on the configuration space used. For example, for particles in motion, properties can be described in terms of coordinates while in electromagnetism, L could be expressed in terms of fields and potentials. We thus use the general coordinate q_i and express the action J characterizing the system as:

$$J = \int_{t_1}^{t_2} L(q_i, \dot{q}_i, t)dt \tag{3.44}$$

We note that the integration parameter is time, t, and $\dot{q}_i$ stands for a generalized velocity, $\frac{\partial q_i}{\partial t}$. We want this action to be extremal. This looks very much like the situation we encountered earlier and solved by means of the calculous of variations technique. The difference is in the choice of the coordinates and of the integration parameter, which is now time. We analyze Eq. 3.44 in a similar way. We write the effect of a small change ε on coordinate q_i as:

$$q(t, \alpha) = q_i(t, 0) + \varepsilon\eta(q_i) \tag{3.45}$$

where $\eta(q_i)$ must be zero at the two points shown in Figure. 3.5 The integral becomes a function of the small change ε that we introduce in the coordinate q_i. We can thus write:

$$J(\varepsilon) = \int_{t_1}^{t_2} L(q_i(t, \varepsilon), \dot{q}_i(t, \varepsilon), t)\, dt \tag{3.46}$$

We assume that $J(\varepsilon)$ varies continuously with ε. To find that extremum, we differentiate $J(\varepsilon)$ relative to ε, and then equate the result to 0. We differentiate under the integral sign using standard rules of partial differentiation. We thus have:

$$\frac{\partial J(\varepsilon)}{\partial \varepsilon} = \int_{t_1}^{t_2} \left(\frac{\partial L}{\partial q_i} \frac{\partial q_i}{\partial \varepsilon} + \frac{\partial L}{\partial \dot{q}_i} \frac{\partial \dot{q}_i}{\partial \varepsilon} \right) dt \tag{3.47}$$

We have kept t constant, independent of ε as mentioned above, changing only q_i by means of ε. The second term under the integral sign may be written as:

$$\int_{t_1}^{t_2} \frac{\partial L}{\partial \dot{q}_i} \frac{\partial \dot{q}_i}{\partial \varepsilon} dt = \int_{x_1}^{x_2} \frac{\partial L}{\partial \dot{q}_i} \frac{\partial^2 q_i}{\partial t \partial \varepsilon} dt \tag{3.48}$$

We integrate by part this second term and we obtain:

$$\int_{x_1}^{x_2} \frac{\partial L}{\partial \dot{q}_i} \frac{\partial^2 q_i}{\partial t \partial \varepsilon} dt = \left. \frac{\partial L}{\partial \dot{q}_i} \frac{\partial q_i}{\partial \varepsilon} \right|_{t_1}^{t_2} - \int_{x_1}^{x_2} \frac{d}{dt} \left(\frac{\partial L}{\partial \dot{q}_i} \right) \frac{\partial q_i}{\partial \varepsilon} dt \tag{3.49}$$

However, a condition imposed on our calculation is that all possible paths, functions of ε, must pass through the two points at (t_1) and (t_2). Consequently, the first term on the right-hand side of the integral, evaluated at t_1 and t_2, is zero since we require that q_i does not vary with ε at those points and $\left(\frac{\partial q_i}{\partial \varepsilon} \right)_{t_{1,2}} = \eta(t_1, t_2)$ is 0. Eq. 3.47 thus becomes:

$$\frac{\partial J(\varepsilon)}{\partial \varepsilon} = \int_{x_1}^{x_2} \left(\frac{\partial L}{\partial q_i} - \frac{d}{dt} \frac{\partial L}{\partial \dot{q}_i} \right) \frac{\partial q_i}{\partial \varepsilon} dt \tag{3.50}$$

On the other hand, a small change, δJ in J and δq_i in q_i, as a function of ε, evaluated at ε = 0, can be written as:

$$\delta J = \left(\frac{\partial J}{\partial \varepsilon}\right)_0 d\varepsilon \qquad \delta q_i = \left(\frac{\partial q_i}{\partial \varepsilon}\right)_0 d\varepsilon \tag{3.51}$$

We multiply both sides of Eq. 3.50 by $d\varepsilon$ and use these definitions. Since J must be an extremum, it must not change in first order with ε at $\varepsilon = 0$ and consequently $\delta J = 0$.:

$$\delta J = \int_{x_1}^{x_2} \left(\frac{\partial L}{\partial q_i} - \frac{d}{dt}\frac{\partial L}{\partial \dot{q}_i}\right)\delta q_i dt = 0 \tag{3.52}$$

However, δq_i is arbitrary and thus to satisfy that condition, the term in parentheses within the integral must be 0. The final result is thus:

$$\frac{\partial L}{\partial q_i} - \frac{d}{dt}\frac{\partial L}{\partial \dot{q}_i} = 0 \tag{3.53}$$

This differential equation is actually a set of equations that applies to all coordinates q_i. This expression is similar to that obtained for the condition imposed on the function used to find the shortest path between two points on a surface by means of the calculous of variations. This is not surprising. We have used a very similar approach here. We recall also that we did not have to define L. What does the result mean then? We have obtained a set of n differential equations (n generalized coordinates q_i, or degrees of freedom). The set of equations is called *Euler-Lagrange's equations.* To simplify writing we will use the notation *Lagrange's equation*, meaning the set of Eqs. 3.53. We thus have, for each coordinate, an equation that fixes conditions on L for the least action principle to be obeyed. However, we know how the universe or objects in it behave: their evolution obeys Newton's principles. Consequently, when we analyze a situation with a force applied to an object or a system, the Lagrangian in Eq. 3.53 must be selected in order to give the same results as those obtained by means of Newton's principles. The selection of the appropriate Lagrangian is thus a first and major task. In the simplest mechanical cases to be examined below, the Lagrangian will be T-V. However, there are many situations, for example those treating with fields, where the Lagrangian takes a very different form. It may be said that the choice of the appropriate Lagrangian is not evident and is sometimes justified *a posteriori*. What does that prove? We can only say that the whole process is coherent and that if the Lagrangian is properly chosen, the principle of least action is justified.

There is, however another approach that leads to the same set of equations and appears to justify the choice of the Lagrangian defined by means of Eq. 3.41: it is based on a property of Newton's second law, a property that we have introduced earlier and that leads to a principle called d'Alembert's principle. The term (T-V) comes out naturally from the analysis. Let us look at that calculation.

C. D'ALEMBERT'S PRINCIPLE

A first observation that we can make in analyzing the state of the universe is that it seems to be in equilibrium at all times or at least seems to be looking for such a state. What do we mean by equilibrium? It may be described as a situation for which all forces present cancel each other. We may also qualify that situation as stationary. We have seen such a situation in our analysis of the orbital motion of a planet around a body with a large mass, a star, for example. A potential well with a minimum energy is created by the combination of the gravitational field of that large mass and the motion of the smaller mass object. We concluded that stable orbits could be created if the small mass was going at the proper speed. This kind of potential well is illustrated in Figure 3.6a, in a 3-D reference frame.

At the bottom of the well, the potential energy of the object is minimum. In the orbital motion used as an example, the object is at rest in a potential well created by the gravitational interaction of the large and small masses and the angular momentum of the small mass. We concluded also that the situation was very similar to a condition where an equilibrium existed between various forces in action. We note that at the bottom of the well where the object is at a minimum of energy, all forces cancel each other and there is an equilibrium at that point. If we vary the position of the object by an infinitesimal virtual quantity δr in any direction from the minimum, nothing changes in first order.

The bottom of the well is flat in first order. In fact, we can write for a first order virtual displacement δr, independent of time:

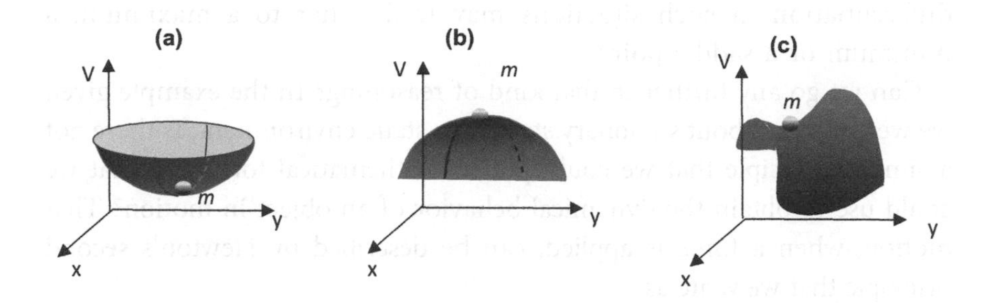

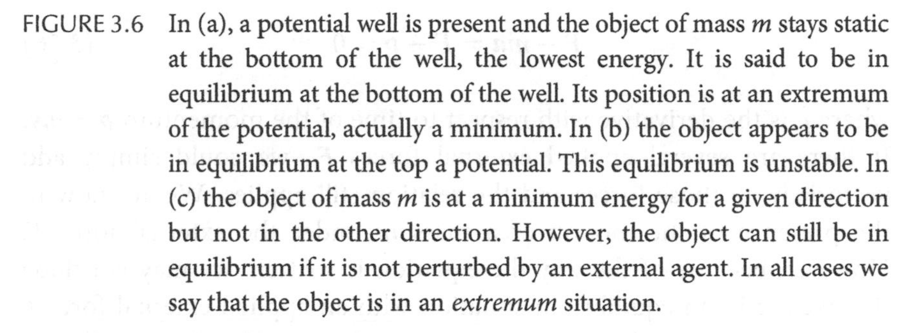

FIGURE 3.6 In (a), a potential well is present and the object of mass m stays static at the bottom of the well, the lowest energy. It is said to be in equilibrium at the bottom of the well. Its position is at an extremum of the potential, actually a minimum. In (b) the object appears to be in equilibrium at the top a potential. This equilibrium is unstable. In (c) the object of mass m is at a minimum energy for a given direction but not in the other direction. However, the object can still be in equilibrium if it is not perturbed by an external agent. In all cases we say that the object is in an *extremum* situation.

$$\delta V = \frac{dV_{object}}{dr} \cdot \delta r = 0 \tag{3.54}$$

Since, for a conservative case, the force can be derived from the gradient of the potential, this can be described as an equilibrium. On the other hand, the product of a force by a small virtual displacement may be identified as virtual work, $\boldsymbol{F} \cdot \delta \boldsymbol{r}$. In fact, the virtual work of all forces applied vanishes because of equilibrium:

$$\sum_i \boldsymbol{F_i} \cdot \delta \boldsymbol{r_i} = 0 \tag{3.55}$$

We note, however, that equilibrium exists also for the other situations illustrated in Figure 3.5b and 3.5c, in which cases all forces also cancel each other. In those cases, we have a maximum for all directions in the case of Figure 3.6b and at least for one direction in the case of the saddle shown in Figure 3.6c. The equilibrium is unstable, however. It is better to qualify such situations as "extrema" and, in general, differentiation in such situations may lead either to a maximum, a minimum or a saddle point.

Can we go any further in that kind of reasoning? In the example given we were talking about stationary states in a static environment. Is there not a similar principle that we could put in mathematical form and that we could use to obtain the dynamical behavior of an object in motion? That motion, when a force is applied, can be described by Newton's second principle that we write as:

$$\boldsymbol{F} - m\boldsymbol{a} = \boldsymbol{F} - \dot{\boldsymbol{p}} = 0 \tag{3.56}$$

where $\dot{p}$ is the derivative with respect to time of the momentum $\boldsymbol{p} = m\mathbf{v}$. If there are several applied external forces $\boldsymbol{F}_i$, we could simply add vectorially all those forces and the relation still applies. We are now in the presence of the motion of a mass m under the external force F. However, because of the way the equation is written, we may conclude that we are in an equilibrium situation with an applied external force F and a reactive force that we may identify as inertia. The situation is very much like a static case. At all times, during the application of the external force, there is equilibrium between the two forces. Consequently, we can introduce the concept of a virtual displacement $\delta\boldsymbol{r}$ as we did earlier for the static case and write:

$$(\boldsymbol{F} - \dot{\boldsymbol{p}}) \cdot \delta\boldsymbol{r} = 0 \tag{3.57}$$

This is known as *d'Alembert's principle.* Eq. 3.57 is interesting, but being in vector form, it does not lead to easy analysis of the general situation of an object being accelerated by a force. However, interpretation of the various terms in Eq. 3.57 and the use of a somewhat moderately sophisticated differential calculus analysis, as we did in the case of the calculus of variations, leads directly to Lagrange's equation, which makes explicit the physical conditions that must be obeyed by the system. Let us do that calculation.

We could use Cartesian or polar coordinates to analyze the equilibrium condition given by Eq. 3.57. However, we have introduced the concept of generalized coordinates and used it earlier to derive Lagrange's equations. We will again use generalized coordinates. We thus write:

$$r = r(q_i, t) \tag{3.58}$$

Using the standard differential analysis techniques developed earlier, we first note that we can express the virtual displacement δr as a function of the coordinates:

$$\delta \boldsymbol{r} = \sum_i \frac{\partial \boldsymbol{r}}{\partial q_i} \delta q_i \tag{3.59}$$

On the other hand, we can write the velocity entering in momentum definition $\boldsymbol{p}$ as:

$$\mathbf{v} = \sum_i \frac{\partial \boldsymbol{r}}{\partial q_i} \frac{dq_i}{dt} + \frac{\partial \boldsymbol{r}}{\partial t} = \sum_i \frac{\partial \boldsymbol{r}}{\partial q_i} \dot{q}_i + \frac{\partial \boldsymbol{r}}{\partial t} \tag{3.60}$$

We can thus rewrite in Eq. 3.57 in terms of these differentials. The first term, $\boldsymbol{F} \cdot \boldsymbol{dr}$, can be written as:

$$\boldsymbol{F} \cdot \delta \boldsymbol{r} = \sum_i F \cdot \frac{\partial \boldsymbol{r}}{\partial q_i} \delta q_i = \sum_i Q_i \delta q_i \tag{3.61}$$

where Q_i was defined earlier and can be considered as the generalized force or the component of the applied force in the coordinate direction q_i:

$$Q_i = \boldsymbol{F} \cdot \frac{\partial \boldsymbol{r}}{\partial q_i} \tag{3.62}$$

In the case of a conservative system, the force may be expressed as the gradient of a potential V

$$\boldsymbol{F} = -\boldsymbol{\nabla} V = -\frac{\partial V}{\partial r} \tag{3.63}$$

Making use of Eq. 3.62 we have:

$$Q_i = -\frac{\partial V}{\partial q_i} \tag{3.64}$$

called the *generalized force component.* On the other hand, the second term of Eq. 3.57 is a little trickier to handle. We have $\dot{\boldsymbol{p}} \cdot \delta r$, which may be written as:

$$\dot{\boldsymbol{p}} \cdot \delta \boldsymbol{r} = m\ddot{\boldsymbol{r}} \cdot \delta \boldsymbol{r} = \sum_j m\ddot{\boldsymbol{r}} \cdot \frac{\partial \boldsymbol{r}}{\partial q_i} \delta q_i \tag{3.65}$$

At this point, there is some algebra to do. Let us expand the term on the left in the following expression:

$$m\frac{d}{dt}\left(\dot{\boldsymbol{r}} \cdot \frac{\partial \boldsymbol{r}}{\partial q_i}\right) = m\ddot{\boldsymbol{r}} \cdot \frac{\partial \boldsymbol{r}}{\partial q_i} + m\dot{\boldsymbol{r}} \cdot \frac{\partial}{\partial q_i}\left(\frac{d\boldsymbol{r}}{dt}\right) \tag{3.66}$$

This can be rearranged as:

$$m\ddot{\boldsymbol{r}} \cdot \frac{\partial \boldsymbol{r}}{\partial q_i} = m\frac{d}{dt}\left(\dot{\boldsymbol{r}} \cdot \frac{\partial \boldsymbol{r}}{\partial q_i}\right) - m\dot{\boldsymbol{r}} \cdot \frac{\partial}{\partial q_i}\left(\frac{d\boldsymbol{r}}{dt}\right) \tag{3.67}$$

We now use Eq. 3.60 and take the derivative of $\mathbf{v}$ relative to the component $\dot{q}_i$, considering it as an independent variable and we obtain:

$$\frac{\partial \mathbf{v}}{\partial \dot{q}_i} = \frac{\partial \boldsymbol{r}}{\partial q_i} \text{ with } \mathbf{v} = \dot{\boldsymbol{r}} = \frac{d\boldsymbol{r}}{dt} \tag{3.68}$$

We replace in Eq. 3.67 and obtain:

$$m\ddot{\boldsymbol{r}} \cdot \frac{\partial \boldsymbol{r}}{\partial q_i} = m\frac{d}{dt}\left(\mathbf{v} \cdot \frac{\partial \mathbf{v}}{\partial \dot{q}_i}\right) - m\dot{\boldsymbol{r}} \cdot \frac{\partial}{\partial q_i}\mathbf{v} \tag{3.69}$$

This can also be written as:

$$m\ddot{r}\cdot\frac{\partial r}{\partial q_i}=\frac{d}{dt}\frac{\partial}{\partial \dot{q}_i}\left(\frac{1}{2}mv^2\right)-\frac{\partial}{\partial q_i}\frac{1}{2}mv^2 \tag{3.70}$$

Thus Eq. 3.57 becomes:

$$\sum_i\left[Q_i-\left\{\frac{d}{dt}\frac{\partial}{\partial \dot{q}_i}\left(\frac{1}{2}mv^2\right)-\frac{\partial}{\partial q_i}\left(\frac{1}{2}mv^2\right)\right\}\right]\delta q_i=0 \tag{3.71}$$

and using Eq. 3.64, we can write:

$$\sum_i\left[-\frac{\partial V}{\partial q_i}-\left\{\frac{d}{dt}\frac{\partial}{\partial \dot{q}_i}\left(\frac{1}{2}mv^2\right)-\frac{\partial}{\partial q_i}\left(\frac{1}{2}mv^2\right)\right\}\right]\delta q_i=0 \tag{3.72}$$

The first term, containing the potential V, can be incorporated under the derivative of the last term. On the other hand, since we assume the system is conservative, the potential is independent of the velocity. We assume also that V is not dependent on time explicitly. Consequently, V can also be subtracted from $\frac{1}{2}mv^2$ in the second term of the equation without altering it and we obtain:

$$\sum_j\left\{\frac{d}{dt}\frac{\partial}{\partial \dot{q}_i}\left(\frac{1}{2}mv^2-V\right)-\frac{\partial}{\partial q_i}\left(\frac{1}{2}mv^2-V\right)\right\}\delta q_i=0 \tag{3.73}$$

The term $\frac{1}{2}mv^2$, as we have seen earlier, is the kinetic energy T of mass m accelerated to velocity $\mathbf{v}$ by the applied force F. Consequently, the term $(T-V)$, which was introduced arbitrarily earlier and that we called the Lagrangian L, appears naturally in Eq. 3.73, which can now be written as:

$$\sum_i\left\{\frac{d}{dt}\frac{\partial}{\partial \dot{q}_i}L-\frac{\partial}{\partial q_i}L\right\}\delta q_i=0 \tag{3.74}$$

That equation is valid for all components of the coordinate system chosen, coordinates that are independent of each other. Consequently, that equation is valid for arbitrary δq_i and the term between brackets must be equal to zero for each of those coordinates. We thus have:

$$\frac{d}{dt}\frac{\partial L}{\partial \dot{q}_i} - \frac{\partial L}{\partial q_i} = 0 \tag{3.75}$$

This equation applies to all coordinates. It is the same differential equation as the one derived earlier by means of a calculation based on the principle of least action with the action defined arbitrarily in terms of the Lagrangian, and which we called Lagrange's equation. Here, however, we have made our analysis by situating ourselves in a frame of reference in which the state of the system, in the context of Newton's second principle, is interpreted as if it were stationary.

We need to reflect on the results obtained. We have performed calculations based on two principles: least action and d'Alembert's principles. We obtained the same results. d'Alembert's principle leads, somewhat indirectly, to an equation identical to that obtained by assuming the least action principle to be valid. It contains a term that is the difference between kinetic and potential energy. This appears to provide a basis for the use of the Lagrangian defined that way in the definition of action and the validation of the principle of least action. Thus, the analysis seems to be coherent. However, physical theories are valid if their predictions are in agreement with observations. In mechanics, we can say that the theory is correct if it agrees with Newton's principles since they have been shown to be correct in general. We will show that this is so in the next chapter. Consequently, it appears that there is no new physics introduced. All was contained already in Newton's principles and equation 3.75 may be considered essentially as the equation of motion. Actually, one could then ask immediately: what's the point? We find, however, that the analysis of the behavior of a mechanical system is reduced to one using scalars rather than vectors, in many cases a great simplification. We only need to postulate a proper Lagrangian and solve differential equation 3.75. We will also see in the next chapter that the approach leads to new interpretation of nature relative to conservation laws and the role of symmetry. Furthermore, we will see that we can extend the analysis to fields such as electromagnetic fields and matter waves. Finally, we will outline its role in a new interpretation of paths in quantum physics, providing a link between classical and quantum mechanics.

We would like to add another comment on this subject. It appears that in our derivation based on the least action principle the object behavior is determined as a whole: it is the full path or action that is minimized. The object seems to know the whole story at the start of its evolution. When d'Alembert's principle is used, it seems that the object sniffs its way along the path, attempting to keep equilibrium. The final result is the same: it leads to Lagrange's equation. Both approaches and behaviors are rather different. However, that observation may simply be a consequence of our lack of understanding of the functioning of the universe, a functioning that we interpret in terms of our mathematical language, which, although logical, may not be the language used by the universe.

CHAPTER IV

Selected Applications of Lagrange's Equations

In this chapter, we will apply Lagrange's equation to various situations. We note that the equation applies to all dynamical variables involved and consists of several equations. We will need to make explicit those dynamical variables and identify the Lagrangian for the various situations chosen. This may not be a simple task. In mechanical systems, the choice of the Lagrangian is normally straightforward. Lagrange's equation may be interpreted essentially as the equation of motion and the chosen Lagrangian should lead to the same results as those obtained using Newton's principles. In other cases, such as in electromagnetism, the choice is not evident and we will see that the calculation is not as intuitive as in the mechanics case.

A. A PARTICLE IN MOTION IN A POTENTIAL

Let us first consider the case of a particle in motion in a potential V that depends on position only. The Lagrangian chosen is the difference between kinetic and potential energy, $L = T - V$. Thus, we have Lagrange's equation for generalized coordinates q_i:

$$\frac{d}{dt}\frac{\partial(T-V)}{\partial\dot{q}_i} - \frac{\partial(T-V)}{\partial q_i} = 0 \tag{4.1}$$

We assume that the potential V does not depend on velocity and the partial derivative in respect to $\dot{q}_i$ is evaluated as:

$$\frac{\partial L}{\partial \dot{q}_i} = \frac{\partial T}{\partial \dot{q}_i} = \frac{\partial}{\partial \dot{q}_i} \sum_j \frac{1}{2} m (\dot{q}_i)^2 = m\dot{q}_i \tag{4.2}$$

Thus:

$$\frac{\partial L}{\partial \dot{q}_i} = p_i \tag{4.3}$$

This term is called the generalized momentum associated with coordinate q_i. The first term in Lagrange's equation is thus reduced to:

$$\frac{d}{dt}\frac{\partial L}{\partial \dot{q}_i} = \dot{p}_i \tag{4.4}$$

On the other hand, the second term reduces to the derivative of V with respect to q_i and according to Eq. 3.64 we identify it as the generalized force:

$$\frac{\partial L}{\partial q_i} = \frac{\partial V}{\partial q_i} = -Q_i \tag{4.5}$$

We thus finally obtain:

$$\frac{\partial L}{\partial q_i} = \dot{p}_i = Q_i \tag{4.6}$$

This is Newton's second dynamical law for coordinate q_i. We may mention that it is what we expected since, in the derivation of Lagrange's equation based on d'Alembert's principle, we essentially started the calculation with that principle. In the case of the derivation using the least action principle the second law was also introduced through the use of the kinetic energy that enters into the Lagrangian. However, we can add that such a result confirms the validity of the least action principle. Consequently, at this stage we can say that Lagrange's equation does not introduce any new physics but is coherent with our understanding of Newton's second law and validates the least action principle.

B. GRAVITATION

Vertical Motion

As another simple example, we can directly derive the laws of vertical motion of an object in a gravitational field using Lagrange's equation. We assume the vertical Cartesian coordinate to be z. There is thus only one degree of freedom and z is used as the generalized coordinate. The system is shown in Figure 4.1, with height h equivalent to z.

The velocity is $\mathrm{v} = \dot{q}_i = \dot{z}$. The kinetic energy is $T = \frac{1}{2}m\mathrm{v}^2 = \frac{1}{2}m\dot{z}^2$ and potential energy near the surface of the earth is $V = mgz$. We have thus added a constant to the potential energy to displace its zero to the surface of the earth. The equation for V is valid only close to that surface. The Lagrangian is thus:

$$L = \frac{1}{2}m\dot{z}^2 - mgz \tag{4.7}$$

We apply Lagrange's equation written for that particular coordinate as:

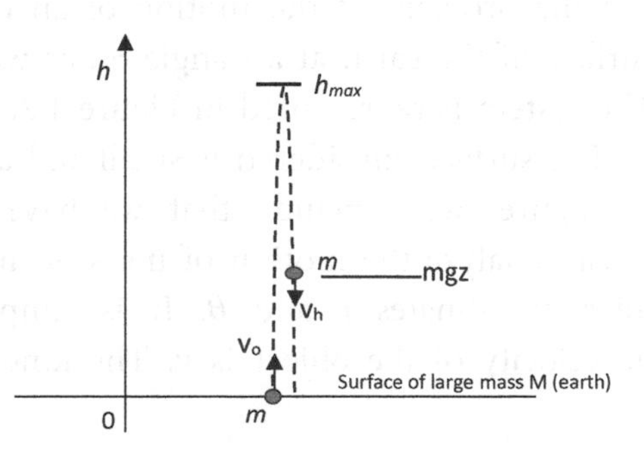

FIGURE 4.1 Motion of a small object of mass m sent upward in the gravitational field of a large object of mass M (earth). The surface of the large object is assumed to be flat.

$$\frac{d}{dt}\frac{\partial L}{\partial \dot{z}} - \frac{\partial L}{\partial z} = 0 \tag{4.8}$$

We obtain:

$$\frac{d}{dt}m\text{v} = -mg \tag{4.9}$$

Cancelling m on each side of the equation and integrating, we obtain

$$\text{v} = \text{v}(0) - gt \tag{4.10}$$

and finally, after a second integration:

$$z(t) = z\ (0) + \text{v}(0)t - \frac{1}{2}gt^2 \tag{4.11}$$

as it should for an object thrown vertically from z = 0 at the earth's surface with an initial vertical velocity v(0). It is observed that the calculation is straightforward since it uses only energy concepts and doesn't not involve vectors. That equation gives of course the same result as that obtained earlier using Newton's principles.

Motion in a Plane Vertical to the Earth's Surface

We now examine the problem of the motion of an object of mass m sent from the surface of the earth at an angle θ_o, as was studied earlier in Chapter II. The system is represented in Figure 4.2, which is adapted from Chapter II. The surface considered is small and assumed flat.

Inspecting the figure, we conclude that we have two degrees of freedom and we can analyze the motion of mass m in either Cartesian x and z or polar coordinates r and θ. It is simpler in Cartesian coordinates. The velocity of the object is v. The kinetic energy of the object is:

$$T = \frac{1}{2}m\text{v}^2 \tag{4.12}$$

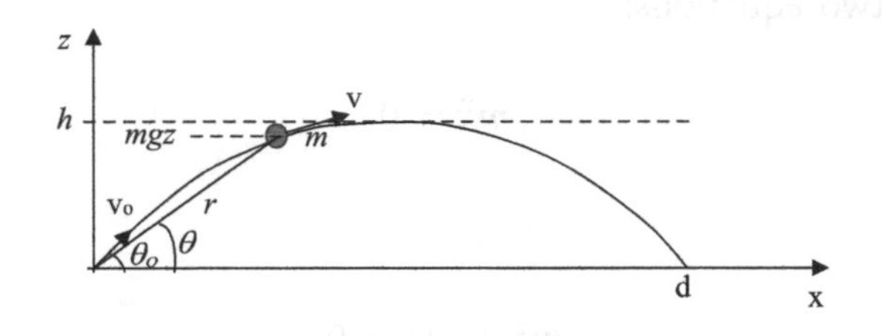

FIGURE 4.2 Trajectory of an object of mass m sent at an angle θ_o from the surface of the earth, assumed flat.

and the velocity v is given by:

$$v^2 = v_x^2 + v_z^2 \tag{4.13}$$

On the other hand, the potential energy of mass m is function of z only. We make the approximation that the height h is much smaller that the radius of the earth. The potential energy is then approximatively as above:

$$V = mgz \tag{4.14}$$

Consequently, the Lagrangian is:

$$L = T - V = \frac{1}{2}m\left(v_x^2 + v_z^2\right) - mgz \tag{4.15}$$

We apply Lagrange's equations to the two coordinates x and z:

$$\frac{d}{dt}\frac{\partial L}{\partial \dot{x}} - \frac{\partial L}{\partial x} = 0 \tag{4.16}$$

and

$$\frac{d}{dt}\frac{\partial L}{\partial \dot{z}} - \frac{\partial L}{\partial z} = 0 \tag{4.17}$$

We obtain two equations:

$$m\ddot{x} = 0 \tag{4.18}$$

and

$$m\ddot{z} + mg = 0 \tag{4.19}$$

Both equations 4.18 and 4.19 are readily integrated to give:

$$x = \mathrm{v}_{ox}t \tag{4.20}$$

$$z = \dot{z}_o t - \frac{1}{2}gt^2 \tag{4.21}$$

Combining these two equations gives the trajectory as a parabola:

$$z = \frac{\mathrm{v}_{zo}}{\mathrm{v}_{xo}}x - \frac{1}{2\mathrm{v}_{xo}}gx^2 \tag{4.22}$$

This is a rather simple calculation and it gives the same result as the one obtained in Chapter II.

Orbital Motion

Let us now examine the question of orbital motion as we did in Chapter II using Newton's principles and the concept of angular momentum, but now using the Lagrangian approach. Assume, as we did, that we have a small object of mass m orbiting a much larger object of mass M. For simplicity we assume the large mass to be so large as not to be affected by the presence of the small mass. The system is shown in Figure 4.3.

We assume that the motion of the small mass takes place in the x-y plane. The angle θ is $\pi/2$ and does not enter into the calculation. From the figure the velocity of the object is given by:

$$\mathrm{v}^2 = \mathrm{v}_r^2 + \mathrm{v}_\varphi^2 \tag{4.23}$$

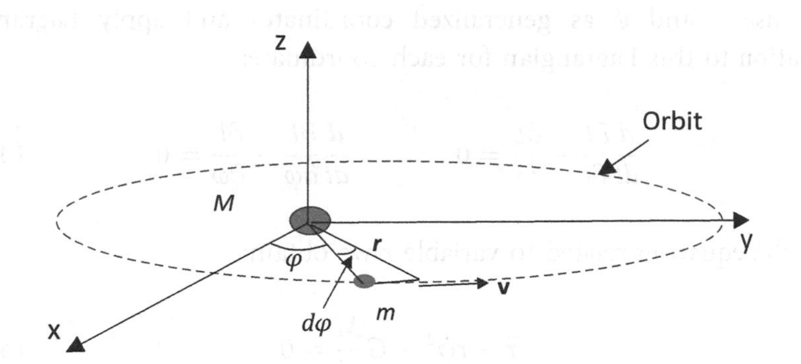

FIGURE 4.3 Motion of a small object of mass m in orbit around an object of large mass M.

We also have:

$$v_r^2 = \left(\frac{dr}{dt}\right)^2 = \dot{r}^2 \tag{4.24}$$

$$v_\varphi^2 = r\dot{\varphi}^2 \tag{4.25}$$

The kinetic energy is thus:

$$T = \frac{1}{2}m\left(\dot{r}^2 + r^2\dot{\varphi}^2\right) \tag{4.26}$$

On the other hand, the potential energy is given by:

$$V = -G\frac{Mm}{r} \tag{4.27}$$

We have the two members to be used in the Lagrangian, which is thus:

$$L = T - V = T = \frac{1}{2}m\left(\dot{r}^2 + r^2\dot{\varphi}^2\right) + G\frac{Mm}{r} \tag{4.28}$$

We use r and ϕ as generalized coordinates and apply Lagrange's equation to this Lagrangian for each coordinate:

$$\frac{d}{dt}\frac{\partial L}{\partial \dot{r}} - \frac{\partial L}{\partial r} = 0 \qquad \frac{d}{dt}\frac{\partial L}{\partial \dot{\varphi}} - \frac{\partial L}{\partial \varphi} = 0 \tag{4.29}$$

For the equation related to variable r we obtain:

$$\ddot{r} - r\dot{\varphi}^2 + G\frac{M}{r^2} = 0 \tag{4.30}$$

In the case of a central force such as that of gravity independent of φ, Lagrange equation reduces to:

$$\frac{d}{dt}\frac{dL}{d\dot{\varphi}} = \frac{d}{dt}\left(mr^2\dot{\varphi}\right) = 0 \tag{4.31}$$

These equations give the relation between radial speed, tangential speed, radial acceleration and tangential acceleration for orbital paths. Eq. 4.31 forces the term in brackets under the time derivative to be constant with time. It is recognized as the angular momentum ℓ defined earlier and it is concluded that, in the case of a central force, angular momentum is conserved. We will expand on this below. On the other hand, the term $(1/2)r^2 d\varphi/dt$ is recognized as the area of the triangle swept in time dt. This area is constant, and this is recognized as one of Keppler's laws for orbital motion.

A circular orbit is easily characterized using Eq. 4.30. In that case there is no radial acceleration and $\ddot{r} = 0$. The velocity, being tangential to the circular motion, is equal to:

$$\mathrm{v} = r\dot{\varphi} \tag{4.32}$$

This leads to:

$$r\dot{\varphi}^2 = \frac{r^2\dot{\varphi}^2}{r} = \frac{\mathrm{v}^2}{r} \tag{4.33}$$

Using Eq. 4.30 we obtain:

$$v = \sqrt{\frac{GM}{r}} \tag{4.34}$$

This equation sets the relation between the orbit radius and the tangential velocity to obtain a circular orbit, as was found earlier in Chapter II (Eq. 2.147). Finally, Eq. 4.30 can reproduce the results obtained in Chapter II describing the orbital motion as that of a single object of mass m in an effective one-dimensional potential well.

C. THE PENDULUM

In Chapter II we analyzed briefly the main characteristics of the pendulum and introduced the concept of generalized coordinates with a constraint. We also analyze the system within the classical concept of conservation of energy using Newton's principles. We can now make that analysis by means of Lagrange's equation with the single coordinate θ. We refer to Figure 2.13, which we reproduce here for convenience as Figure 4.4.

We need to define the Lagrangian for the system. For this we need to evaluate the kinetic energy and the potential energy. We already did that in Chapter II. From the results obtained, we thus have:

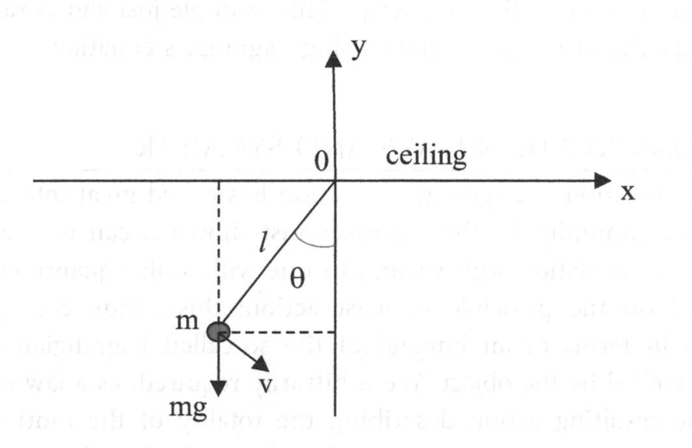

FIGURE 4.4 The simple pendulum.

$$L(\theta,\dot{\theta},t) = T - V = \frac{1}{2}ml^2\dot{\theta}^2 + mglcos\theta \tag{4.35}$$

We use θ as a generalized coordinate. Lagrange's equation for that coordinate is:

$$\frac{d}{dt}\frac{\partial L}{\partial\dot{\theta}} - \frac{\partial L}{\partial\theta} = 0 \tag{4.36}$$

Each term is evaluated with L given by Eq.4.35 and we obtain:

$$\ddot{\theta} + \frac{g}{l}\sin\theta = 0 \tag{4.37}$$

At this point we make an approximation. We assume that the pendulum oscillation amplitude is small and we approximate $\sin\theta$ by θ. We thus have:

$$\ddot{\theta} = -\frac{g}{l}\theta \tag{4.38}$$

The solution of that equation is well known as a cosine function with two constants of integration that may be fixed by means of initial conditions. The result is the same as that obtained earlier through the principle of conservation of energy. This example just shows rather well the simplicity of the calculations using Lagrange's equation.

D. CONSERVATION LAWS AND SYMMETRY

The introduction of Lagrange's equation has raised great interest in the physics community. In the examples just shown it can be seen that it reduces a calculation with vectors to one with scalar quantities only. It is based on the principle of least action; this *action* concept being defined in terms of an integral of the so-called Lagrangian over the path travelled by the object. We arbitrarily required, as a law of nature, that the resulting action describing the totality of the motion of the object to be an extremum. We can therefore assimilate that condition as being the result of nature looking for equilibrium at all times. This was clearly made explicit by deriving Lagrange's equation also by means of d'Alembert's principle. At this stage, it is natural to ask if the conditions imposed on the Lagrangian by that equation do not lead to other specific laws. In fact, this subject has been the object of intense theoretical analysis and has led to the identification of fundamental properties of the universe we live in. One of those properties regards the connection of the concept of symmetry to the conservation laws that we enunciated earlier. There is an implicit relation between the two. If we examine carefully Lagrange's equation, we observe that we are in the presence of derivatives of L with respect to variables q_i and $\dot{q}_i$. If the components of L, that is T or V, are independent of one of those variables, the partial derivative relative to that variable is zero and the equation is reduced to one term being equal to zero. On the other hand, it is readily realized that, in general, the existence of a relation of the form:

$$\frac{dX}{dt} = 0 \tag{4.39}$$

leads to the conclusion that the quantity X must be independent of time for the relation to be valid. We thus conclude that the quantity X is a constant with time. We then say that the quantity keeps its properties with time and is *conserved*. We used that property earlier without elaborating much on it.

Conservation of Linear Momentum

We can use this property with the Lagrangian and attempt to deduct the properties of objects in motion directly from Lagrange's equation. We recall that Lagrange's equation was obtained by requiring that the action was an extremum, that is $\delta J = 0$.

Let us consider the case where an object is in motion, with its kinetic energy and potential energy both independent of position q_i. We also assume that the applied force is conservative and V is independent of velocity. In that case, all operations involving the derivative with respect to q_i are zero. The variable q_i is then said to be ignorable and, in the jargon associated with this field, it is called a cyclic variable. In that case, Lagrange's equation reduces to one term equal to zero:

$$\frac{d}{dt}\left(\frac{\partial L}{\partial \dot{q}_i}\right) = 0 \tag{4.40}$$

As we just said, this implies that the term under the derivative with respect to time must be a constant for the equation to be valid. However, we have shown that this term is equal to the generalized momentum (Eq. 4.3) and we can write:

$$\frac{d}{dt} p_i = 0 \tag{4.41}$$

We thus conclude from the previous argument that

$$p_i = constant \tag{4.42}$$

Since the coordinate chosen was arbitrary, this result is thus general for all components of p. We conclude that the generalized momentum in the case of an ignorable or cyclic coordinate is conserved.

We have just shown that, if V is independent of q_i, the conjugate momentum is conserved. What happens, however, if we move the whole system along coordinate q_i, by a quantity δq_i as shown in Figure 4.5.

Suppose that we are studying a given characteristic of a system and we find that this characteristic does not change upon that small translation δq_i: we can say that the characteristic is conserved in translation. We can then repeat that small translation and obtain the same result again. This operation finally results in a finite translation, and we can

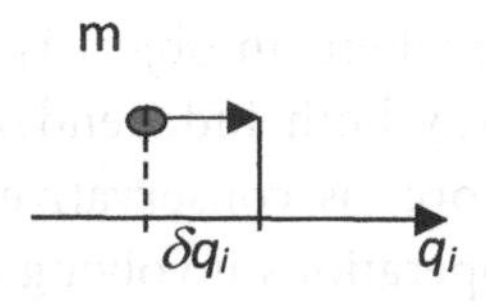

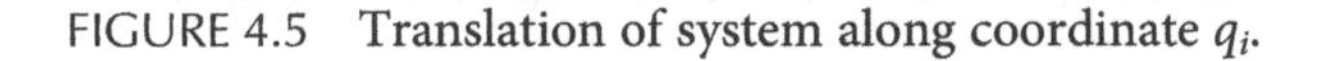

FIGURE 4.5 Translation of system along coordinate q_i.

say that if a given property of the system is not changed by the translation, we have conservation of that property. On the other hand, we are familiar with the concept that if an object keeps a given property or characteristic, when submitted to a given transformation, we say that we are in the presence of a symmetry of the property or characteristic. For example, if in the translation in space of an object with a given volume it is found that the volume stays the same, we say that the volume is a symmetry of the object under translation.

Consequently, it makes sense to ask if the action as defined earlier is symmetrical under translation. The action was defined mathematically as:

$$J = \int_{t_1}^{t_2} L(q_i, \dot{q}_i, t)\, dt \tag{4.43}$$

and is a function of the coordinates q_i and $\dot{q}_i$ Upon translation by δq_i, we note that there could result also a change in velocity by $\delta \dot{q}_i$. The change in J, written as δJ, may then be written as:

$$\delta J = \int_{t_1}^{t_2} \left(\frac{\partial L}{\partial q_i} \delta q_i + \frac{\partial L}{\partial \dot{q}_i} \delta \dot{q}_i \right) dt \tag{4.44}$$

If the coordinate q_i is cyclic, that is ignorable in T and V, the derivative of L with respect to q_i is then zero. Similarly, if we are dealing with a translation without change in velocity we have $d\dot{q}_i$ equal to zero and the second term is also zero. Consequently, in that case $\delta J = 0$ and J is not altered by the translation. It follows that translation is a symmetry of the action. On the other hand, we may expand terms in Eq. 4.44 as follows. We first introduce the following relation obtained by simple differential calculus:

$$\frac{d}{dt}\left(\frac{\partial L}{\partial \dot{q}_i} \delta q_i \right) = \left\{ \frac{d}{dt}\left(\frac{\partial L}{\partial \dot{q}_i} \right) \right\} \delta q_i + \frac{\partial L}{\partial \dot{q}_i} \frac{d}{dt} \delta q_i \tag{4.45}$$

We take the time derivative of the last term on the right-hand side explicit in q_i and rearrange the terms of this last equation to obtain:

$$\frac{\partial L}{\partial \dot{q}_i}\delta \dot{q}_i = \frac{d}{dt}\left(\frac{\partial L}{\partial \dot{q}_i}\delta q_i\right) - \frac{d}{dt}\left(\frac{\partial L}{\partial \dot{q}_i}\right)\delta q_i \tag{4.46}$$

We replace this in Eq.4.43 and obtain:

$$\delta J = \int_{t_1}^{t_2}\left(\frac{\partial L}{\partial q_i}\delta q_i + \frac{d}{dt}\left(\frac{\partial L}{\partial \dot{q}_i}\delta q_i\right) - \frac{d}{dt}\left(\frac{\partial L}{\partial \dot{q}_i}\right)\delta q_i\right)dt \tag{4.47}$$

which we rewrite as:

$$\delta J = \int_{t_1}^{t_2} d\left(\frac{\partial L}{\partial \dot{q}_i}\delta q_i\right) + \int_{t_1}^{t_2}\left(\frac{\partial L}{\partial q_i} - \frac{d}{dt}\left(\frac{\partial L}{\partial \dot{q}_i}\right)\right)dt\ \delta q_i = 0 \tag{4.48}$$

We have set this result equal to zero since in the case of symmetry we have concluded that the action is unaltered by a change in coordinate or by a translation. Furthermore, δq_i being arbitrary, the second integral is zero since we have assumed a conservative force and the term within brackets, which we identify as the left-hand member of Lagrange equation (Eq. 3.75), is equal to zero. The first term on the right-hand side is readily integrated and we obtain:

$$\delta J = \frac{\partial L}{\partial \dot{q}_i}\delta q_i\Big|_{t_1}^{t_2} = 0 \tag{4.49}$$

The term $\frac{\partial L}{\partial \dot{q}_i}$ within the integral is the conjugate momentum p_i and thus we can write:

$$\delta J = p_i\,\delta q_i\Big|_{t_1}^{t_2} = 0 \tag{4.50}$$

The translation δq_i is arbitrary. Consequently, since the integration limits are also arbitrary, the value of the integral must be the same at both limits t_1 and t_2 for the equation to be valid. Thus, p_i must remain constant with translation and the linear momentum is conserved.

Interpreting space symmetry as we did, that is a translation in space by an arbitrary value that leaves the action invariant, we conclude that linear momentum is invariant or is conserved under spatial symmetry.

Conservation of Angular Momentum

We now show that angular momentum is conserved using a similar argument as the one used for the linear momentum. First, let us recall some properties of the concept that we defined earlier as angular momentum. Referring to Figure 2.14, describing the motion of a mass in orbit around a larger mass in the x–y plane, we have defined the angular momentum ℓ as:

Angular momentum = moment of linear momentum

$$\ell = \boldsymbol{r} \times m\mathbf{v} \tag{4.51}$$

We have also defined the torque:

Torque = moment of an external applied force that could alter **v**

$$\boldsymbol{N} = \boldsymbol{r} \times \boldsymbol{F} \tag{4.52}$$

Since $\boldsymbol{F} = \frac{d}{dt} m\mathbf{v}$ and $\boldsymbol{r}$ has a constant length, we thus have the property that

$$\boldsymbol{N} = \boldsymbol{r} \times \frac{d}{dt}(m\mathbf{v}) = \frac{d}{dt}\boldsymbol{r} \times m\mathbf{v} = \frac{d}{dt}\ell \tag{4.53}$$

All these parameters are illustrated in Figure 4.6. Here, we have assumed a force of constraint, for example the attraction of large mass M situated at the origin, which maintains the small mass in circular orbit and is conservative. The force is always at right angle to the motion of the small mass and thus does not change its speed, only its direction. On the other hand, the angular momentum ℓ is the vector product of $\boldsymbol{r}$ and velocity **v** and is thus perpendicular to the x–y plane of the figure. The torque $\boldsymbol{N}$ that could be applied to alter speed v would be in the same direction as ℓ.

The velocity of the object of mass m in the system shown is readily calculated, the radius of the orbit being in the x–y plane with the angle $\theta = \pi/2$, so we have:

$$\mathrm{v} = r\dot{\varphi} \tag{4.54}$$

and the kinetic energy is thus given by:

$$T = \frac{1}{2} m\mathrm{v}^2 = \frac{1}{2} mr^2\dot{\varphi}^2 \tag{4.55}$$

It is independent of coordinate φ. Thus, coordinate φ is ignorable in T. It is also ignorable in V since we have assumed a circular orbit and that no torque is applied from external sources.

The Lagrangian is thus reduced to:

$$L = T - V = \frac{1}{2} mr^2\dot{\varphi}^2 \tag{4.56}$$

Consequently, Lagrange's equation, for variable φ,

$$\frac{d(T-V)}{d\varphi} - \frac{d}{dt}\left(\frac{d(T-V)}{d\dot{\varphi}}\right) = 0 \tag{4.57}$$

reduces to:

$$\frac{d}{dt}\left(\frac{d}{d\dot{\varphi}} \frac{1}{2} mr^2\dot{\varphi}^2\right) = 0 \tag{4.58}$$

Using the same argument as the one used above, the quantity within brackets must be a constant with time for this equation to be valid, and consequently it is conserved. Differentiating with respect to $\dot{\varphi}$, this quantity becomes $mr^2\dot{\varphi}$, and is the angular momentum ℓ of particle with mass m in circular orbit around the object with mass M. Thus, the angular momentum is a constant of motion and is conserved.

However, we can go a step further by asking if that property is maintained when we make a translation of the mass coordinate in its orbit. We requested that, when there is space symmetry, translation would not alter the action or:

$$\delta J = \frac{\partial L}{\partial \dot{q}_i} \delta q_i \bigg|_{t_1}^{t_2} = 0 \tag{4.59}$$

Since the orbit is assumed to be circular, r is a constant and we have:

$$dq_i = r d\varphi \quad \delta q_i = r \delta\varphi \quad d\dot{q}_i = r d\dot{\varphi} \tag{4.60}$$

The change in action is:

$$\delta J = \frac{\partial L}{\partial \dot{\varphi}} \delta\varphi \bigg|_{t_1}^{t_2} = 0 \tag{4.61}$$

which becomes:

$$\frac{\partial L}{\partial \dot{\varphi}} \delta\varphi \bigg|_{t_1}^{t_2} = mr^2 \dot{\varphi} \delta\varphi \big|_{t_1}^{t_2} = 0 \tag{4.62}$$

Since $\delta\varphi$ *is* arbitrary, the term $mr^2\dot{\varphi}$ must be the same at both limits t_1 and t_2 for the integral to be zero. Thus, the angular momentum $mr^2\dot{\varphi}$ stays constant and is conserved in the translation δq_i, which in the present case corresponds to a rotation for which the action does not vary. We thus conclude that symmetry under rotation preserves angular momentum.

We can make another step by assuming that instead of a mass m in orbit around large mass M, we assume that the object is a large mass of a symmetrical shape like a top whose axis of rotation is centered on the z axis. We can assume that the top is made of small masses, particles held together by forces dependent only on the distance between them, not on their absolute position. These forces hold the particles together. If the top is in rotation, all particles experience a movement like the one shown in Figure 4.6, but are situated at different heights on z and different distances from axis z. The analysis made above applies to all

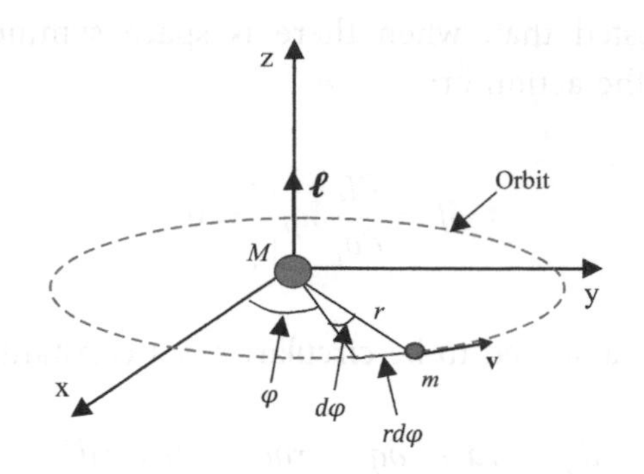

FIGURE 4.6 Simple system used to study conservation of angular momentum.

those particles and the final result needs to be summed over all those particles, their mutual interactions (constraints) not entering into play.

We thus conclude from this exercise that symmetry around the z axis implies conservation of angular momentum, in a similar way as symmetry along an axis implies conservation of linear momentum. These conclusions are valid for all axes and we thus conclude that spatial symmetry implies conservation of linear momentum and spherical symmetry implies conservation of angular momentum.

Conservation of Energy

Earlier we arrived at the conclusion that if applied forces were derived from a so-called conservative potential, energy was conserved. Can we arrive at the same conclusion as a property that follows from the validity and generality of Lagrange's equation? Let us look at that. We have arrived at conclusions on conservation of linear momentum and angular momentum as a result of symmetry with space, that is translation and rotation. Can we do something similar with energy? In that case what will we alter or change? What variable or parameter do we have left? We can make a small translation in time, δt, and analyze its effect on the action J as we did for space translation. However, J is an integral of the Lagrangian L and if L does not change under time translation, we can conclude that J will not change either. When we do a time translation, we assume that we do nothing to space coordinates and velocities. We thus write the transformation as:

$$q_i(t) \to q_i(t') \qquad \dot{q}_i(t) \to \dot{q}_i(t') \qquad where \qquad t' \to t + \delta t \tag{4.63}$$

In that case, a change of L due to a translation of time is present only if L is an explicit function of time:

$$\delta L = \frac{\partial L}{\partial t}\delta t \tag{4.64}$$

However, there can be a time dependence of the q's and $\dot{q}'s$. In that case, considering all time dependences, the total time variation of the Lagrangian, summing over all coordinates q_i, may be written:

$$\frac{dL}{dt} = \sum_i \frac{\partial L}{\partial q_i}\frac{\partial q_i}{\partial t} + \sum_i \frac{\partial L}{\partial \dot{q}_i}\frac{\partial \dot{q}_i}{\partial t} + \frac{\partial L}{\partial t} \tag{4.65}$$

In the following calculation we will assume that L has a time dependence only through the q's and $\dot{q}'s$. Consequently, we assume that L is not an explicit function of time. We also assume that the forces are conservative. In that case, Lagrange's equation applies and we have:

$$\frac{\partial L}{\partial q_i} = \frac{d}{dt}\frac{\partial L}{\partial \dot{q}_i} \tag{4.66}$$

Replacing, in Eq. 4.65, the value of $\frac{\partial L}{\partial q_i}$ given by that equation, we obtain:

$$\frac{dL}{dt} = \sum_i \frac{d}{dt}\frac{\partial L}{\partial \dot{q}_i}\frac{\partial q_i}{\partial t} + \sum_i \frac{\partial L}{\partial \dot{q}_i}\frac{\partial \dot{q}_i}{\partial t} \tag{4.67}$$

or:

$$\frac{dL}{dt} = \sum_i \frac{d}{dt}\frac{\partial L}{\partial \dot{q}_i}\dot{q}_i + \sum_i \frac{\partial L}{\partial \dot{q}_i}\frac{\partial \dot{q}_i}{\partial t} \tag{4.68}$$

which can be written as:

$$\frac{dL}{dt} = \sum_i \frac{d}{dt}\left(\dot{q}_i \frac{\partial L}{\partial \dot{q}_i}\right) \tag{4.69}$$

or

$$\frac{d}{dt}\left(L - \sum_i \dot{q}_i \frac{\partial L}{\partial \dot{q}_i}\right) = 0 \tag{4.70}$$

This means that the quantity in brackets, in the case that L is not an explicit function of time, must be independent of time for the equation to be valid. What is this quantity? It has been called the Hamiltonian, $-H$:

$$\left(L - \sum_i \dot{q}_i \frac{\partial L}{\partial \dot{q}_i}\right) = -H = constant \tag{4.71}$$

The term on the left-hand side may be evaluated as follows. The term $\frac{\partial L}{\partial \dot{q}_i}$ under the integral is the conjugate momentum p_i (Eq. 4.3) and thus we have:

$$L - \sum_i \dot{q}_i p_i = -H \tag{4.72}$$

We interpret readily p_i as $m\dot{q}_i$. The term under the summation is then equal to twice the kinetic energy associated to coordinate q_i. Summing over all coordinates we obtain:

$$L - 2T = T - V - 2T = -T - V = -H \tag{4.73}$$

We thus have:

$$H = T + V = \text{constant} \tag{4.74}$$

This is the total energy and it is the quantity that is constant upon translation of time. We thus conclude that *time symmetry*, represented here by a translation in time, does not alter the Lagrangian and, consequently, the action. It thus leads to the principle of *conservation of energy*.

These conclusions relative to the connection between symmetry and conservation of various quantities such as linear momentum, angular momentum and energy are probably the most profound results to be obtained by this type of analysis done within the Lagrangian formalism. Those particular conservation principles are examples of Noether's theorem, which associates a conservation property to a given symmetry and vice versa. This leads to the general statement that if a given symmetry is found in nature then an associated conservation property law must exist also. This is one of the most profound principles that we can enunciate according to our knowledge of the physics of the universe.

We will now look at some other subjects where the formalism either reduces calculation or helps in understanding the physics involved.

E. HAMILTON'S EQUATIONS OF MOTION

It is possible to obtain, from the previous analysis using Lagrange's equation, new simpler equations that describe the motion of an object by means of considerations on the Hamiltonian just introduced. Equation 4.71 defining the Hamiltonian may be written as:

$$H = \left(\sum_i \dot{q}_i \frac{\partial L}{\partial \dot{q}_i} \right) - L = \sum \dot{q}_i p_i - L \tag{4.75}$$

Where we have used the property:

$$p_i = \frac{\partial L}{\partial \dot{q}_i} \tag{4.76}$$

From Eq. 4.75, we may write the infinitesimal change of H upon infinitesimal changes of variables, p_i, q_i and $\dot{q}_i$:

$$dH = \sum_i \dot{q}_i dp_i + \sum_i p_i d\dot{q}_i - \sum_i \frac{\partial L}{\partial q_i} dq_i - \sum_i \frac{\partial L}{\partial \dot{q}_i} d\dot{q}_i - \frac{\partial L}{\partial t} dt \tag{4.77}$$

We note that the second and fourth terms on the right-hand side of that equation cancel each other. Furthermore, we have:

$$p_i = \frac{\partial L}{\partial q_i} \tag{4.78}$$

and the equation becomes:

$$dH = \sum_i \dot{q}_i dp_i - \sum_i \dot{p}_i dq_i - \frac{\partial L}{\partial t} dt \tag{4.79}$$

But we can also write dH as a full differential in terms of the q, p and t variables. We obtain:

$$dH = \sum_i \frac{\partial H}{\partial p_i} dp_i + \sum_i \frac{\partial H}{\partial q_i} dq_i + \frac{\partial H}{\partial t} dt \tag{4.80}$$

Comparing these two equations we can write the following first order equations:

$$\frac{\partial H}{\partial p_i} = \dot{q}_i \tag{4.81}$$

$$\frac{\partial H}{\partial q_i} = -\dot{p}_i \tag{4.82}$$

and finally:

$$\frac{\partial H}{\partial t} = -\frac{\partial L}{\partial t} \tag{4.83}$$

These are Hamilton's equations. They set, just as Lagrange's equation does, the dynamics of evolution of an object in space with H, the total energy obeying the principle of conservation of energy. They are first order differential equations, and simpler than Lagrange's equation.

F. SPECIAL RELATIVITY

The analysis just made using the least action principle leads us to an understanding of the dynamics of the universe that is rather surprising and most interesting. The universe in its evolution continuously looks for equilibrium and selects the easiest or shortest path in reaching its final state. We can also say that in its evolution the universe maintains equilibrium between acting forces. For example, we have re-written Newton's second principle in such a way that it appears that the force applied to accelerate a particle is continuously facing inertia, which is itself acting as a force opposing the applied force: the moving system seems to be in an equilibrium state at all times or appears to be static. This led us to d'Alembert's principle from which we derived Lagrange's equation. We have also shown that the universe, by the same fact, obeys the least action principle.

However, the world is a little more complex than the one we have looked at. The universe operates in a "relativistic" framework in which it is not possible to determine in an absolute way the velocity of an object. Speed is always relative to some arbitrary point in space, a point that is determined by the observer who may consider himself either moving or at rest. In other words, the motion of one object is relative to another object's motion. There is no such thing as absolute space in which an object could be considered at rest. On the other hand, aside from some motion that can be recognized as local, we observe that galaxies recess from each other. It looks as if the universe expands and as if all matter originated from a single point or a small volume. Outside that original location, which in some model is considered as an infinitesimal point, there is no space and what we call space is in a sense created at the same time as all matter in the universe. We could use galaxies as references of an expanding universal grid and refer all motion to that grid. Since an observer within one of those galaxies may consider himself at rest relative to another galaxy and vice versa, it is clear that the laws of mechanics, enunciated by means of Newton's principles, apply independently equally well in all galaxies. Consequently, Newton's laws apply in any coordinate system considered at rest or in uniform motion.

Another important observation is concerned with the speed of electromagnetic radiation that we call c. As mentioned previously, it was found experimentally that light propagates at the same speed in any coordinate systems in motion relative to each other. Or, in mathematical terms, if two systems are moving relative to each other at velocity **v** and if a pulse of electromagnetic radiation passes by both systems, observers in either system measure the speed of the passing radiation as *c*. Furthermore, no object can travel faster than *c*. In fact, *c*, as determined by means of the theory of electromagnetism, is given by:

$$c = \frac{1}{\sqrt{\varepsilon_o \mu_o}} \tag{4.84}$$

where ε_o and μ_o are electrostatic and magnetostatics constants of the vacuum of space and identified as static properties of vacuum. The constant *c* is equal to 299 792 458 m/s. The meter is now defined as the space travelled by light in (1/299 792 458) second. Since that number refers to static properties of vacuum we may consider that parameter simply as a fundamental constant of nature or universe. That speed appears to be a speed limit characterizing the universe and it happens that light travels at that speed. This observation defies common sense and, as pointed out by Poincaré, appears to be a conspiracy of the universe to prevent measurement of the absolute motion of an observer. We might accept it as fundamental property or a law of nature. This property was exploited by Einstein in the development of the theory of relativity. In that theory, the speed of light is assumed to be a constant whatever the speed of the coordinate system in which the observer or the source are situated. Furthermore, he assumed that all laws of physics are the same in any inertial systems.

We can analyze the effect of the constancy of *c* on coordinate systems. Assume two systems of coordinates whose origins coincide at time $t = 0$ and assume that system Σ' is moving relative to system Σ as shown for simplicity in a 2-D system in Figure 4.7 with speed v along the *x* axis.

A pulse of light radiating from the origin at time t = 0 propagates in both systems at speed *c*. We recall the analysis we made in Chapter II. The equation describing the position of the pulse in system Σ is:

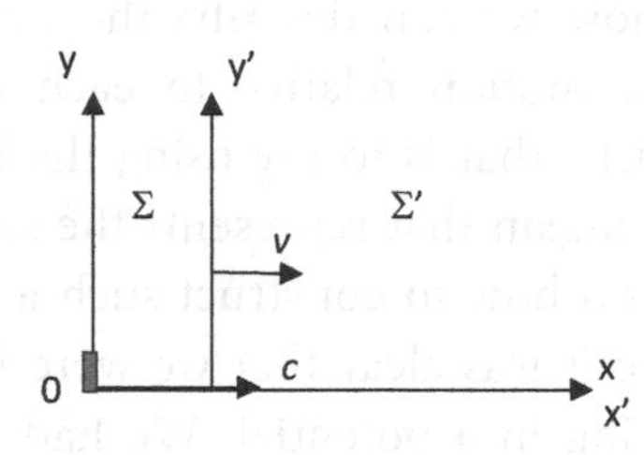

FIGURE 4.7 Two systems of reference moving relative to each other at speed v along the x axis.

$$x^2 + y^2 + z^2 = ct^2 \tag{4.85}$$

Similarly, we have in frame Σ':

$$x'^2 + y'^2 + z'^2 = ct'^2 \tag{4.86}$$

in which we have allowed time to have different properties in both systems. We elaborated earlier on the connection between the coordinates and time and show that we can consider space and time as intimately correlated. In fact, we can consider time as a fourth coordinate and we found unique relations between coordinates of the two systems. The result is:

$$x' = \frac{x - \mathrm{v}t}{\sqrt{1 - \mathrm{v}^2/c^2}} \quad y' = y \quad z' = z \quad t' = \frac{t - \mathrm{v}x/c^2}{\sqrt{1 - \mathrm{v}^2/c^2}} \tag{4.87}$$

These equations show that explicit relations exist between space and time coordinates in systems moving relative to each other. It shows in particular that time does not flow at the same rate in both systems as measured by an observer at rest in one of the two systems. We note immediately that, if the speed v is very small, we have time flowing at the same rate in both systems, the term v^2/c^2 becoming negligible in front of 1 in the denominator of the expressions. We then have, as expected, a Galilean coordinate transformation of x to x' involving only the change "vt" in the x direction. Time remains the same in both systems to first order due to the smallness of (v/c).

We may now ask how we can describe the motion of objects with mass m in systems in motion relative to each other. Following the approach used previously, that is to say using the least action principle, we need to find a Lagrangian that represents the situation. We have not developed a rigid rule on how to construct such a Lagrangian except in the simplest case where it was clear that we were in the presence of an object of mass m moving in a potential. We had a term describing its internal property, that is kinetic energy, and a term describing the effect of external forces through that potential. We required that Lagrange's equation using the chosen Lagrangian verifies Newton's equation of motion. It is not clear, however, what the Lagrangian should look like when relativity is taken into account. It should most certainly contain a term in v/c, which when approximated to low velocity should lead to Newtonian mechanics. From previous intuitive analysis, the following expression is suggested:

$$L(q_i, \dot{q}_i, \mathrm{v}) = -m_0 c^2 \sqrt{1 - \left(\frac{\mathrm{v}}{c}\right)^2} - U(q_i) \tag{4.88}$$

where m_o is the so-called object rest mass to be qualified below, v is the speed of the object equal to $\dot{q}_i$ and c is the speed of light, and $U(q_i)$ is the potential energy. Let us use this expression in Lagrange's equations written as:

$$\frac{dL}{dq_i} = \frac{d}{dt}\frac{\partial L}{\partial \dot{q}_i} \tag{4.89}$$

We evaluate the various terms:

$$\frac{\partial L}{\partial \dot{q}_i} = \frac{m_o \mathrm{v}}{\sqrt{1 - \left(\frac{\mathrm{v}}{c}\right)^2}} \tag{4.90}$$

$$\frac{\partial L}{\partial q_i} = -\frac{\partial U(q_i)}{\partial q_i} = F(q_i) \tag{4.91}$$

where F is the force applied. We thus have:

$$\frac{\mathrm{d}}{\mathrm{d}t}\frac{m_o\dot{q}_i}{\sqrt{1-\left(\frac{\dot{q}_i}{c}\right)^2}} = F(q_i) \tag{4.92}$$

This is essentially Newton's second law of inertia with m now given by:

$$m = \frac{m_o}{\sqrt{1-\left(\frac{\mathrm{v}}{c}\right)^2}} \tag{4.93}$$

Consequently, we appear to be on the right track with the Lagrangian chosen. Let us examine the Hamiltonian, defined earlier, for motion in the direction q_i:

$$H = \dot{q}_i\frac{\partial L}{\partial \dot{q}_i} - L \tag{4.94}$$

Using the Lagrangian written above and identifying v again as $\dot{q}_i$ we obtain:

$$H = \frac{m_o c^2}{\sqrt{1-\frac{\mathrm{v}^2}{c^2}}} + U(q_i) \tag{4.95}$$

For v small relative to c we obtain by expansion:

$$H = m_o c^2 + \frac{1}{2}m_o\mathrm{v}^2 + U(q_i) \tag{4.96}$$

This remains, as introduced earlier, the total energy. We see that it is now written in terms of m_o. That mass is considered as the rest mass of the object. The first term, $m_o c^2$, is thus a new term called the rest energy. Mass, as we said, is energy by itself and kinetic energy is the part of energy associated with the motion of the mass in a given frame of reference.

The momentum can now be written as:

$$(p_i)^2 = \frac{m_o^2}{1-\frac{\mathrm{v}^2}{c^2}}\mathrm{v}^2 \tag{4.97}$$

This can readily be transformed into a more instructive expression. We use an easily verified identity such as:

$$\frac{1}{1-\mathrm{v}^2/c^2}-\frac{\mathrm{v}^2/c^2}{1-\mathrm{v}^2/c^2}=1 \tag{4.98}$$

written as:

$$\gamma^2-\beta^2\gamma^2=1 \tag{4.99}$$

with

$$\beta=\frac{\mathrm{v}}{c} \text{ and } \gamma=\frac{1}{\sqrt{1-\beta^2}} \tag{4.100}$$

Multiply both sides of Eq. 4.99 by $m_o^2c^4$:

$$m_o^2c^4\gamma^2-m_o^2c^4\beta^2\gamma^2=m_o^2c^4 \tag{4.101}$$

The second term on the left-hand side may be written using Eq. 4.97:

$$m_o^2c^4\beta^2\gamma^2=(p_i)^2c^2 \tag{4.102}$$

We thus obtain:

$$m_o^2c^4\gamma^2-(p_i)^2c^2=m_o^2c^4 \tag{4.103}$$

On the other hand, the first term on the left-hand side is energy that is readily identified as relativistic kinetic energy containing rest mass energy. We thus obtain the expression:

$$T^2=p^2c^2+m_o^2c^4 \tag{4.104}$$

The Hamiltonian, being total energy as shown previously, can thus be written:

$$Total\ energy = E = \sqrt{p^2c^2 + m_o^2c^4} + U \tag{4.105}$$

This is an expression for the total energy that is usually derived in the context of special relativity. Here, we have derived it on the basis of the validity of the principle of least action, assuming a particular form for the Lagrangian made explicit by Eq. 4.88.

G. UNIVERSE EXPANSION AND GENERAL RELATIVITY

It is also possible, using Lagrange's equation, to obtain the essential elements of the dynamical equation describing the expansion of the universe. This is done without using Einstein field equations and the result is identical to that obtained by Friedmann in the solution of those equations. On the other hand, Hubble observed experimentally that in general galaxies appear to recess from each other. He concluded that the universe was expanding. An analysis can be made using Poisson's equation describing the behavior of the gravitational field (see, for example, Peebles 1993; Vanier 2011). We wish to do this now using the Lagrangian approach.

We first assume that the distance l between two galaxies as shown in Figure 4.8 increases with time as:

$$l(t) = a(t)l_o \tag{4.106}$$

where l_o is the distance at a given time t_o and $a(t)$ is the coefficient of expansion.

We assume that the galaxies have mass m_1 and m_2 respectively. From the equation above we may write the recession velocity v of the two galaxies as:

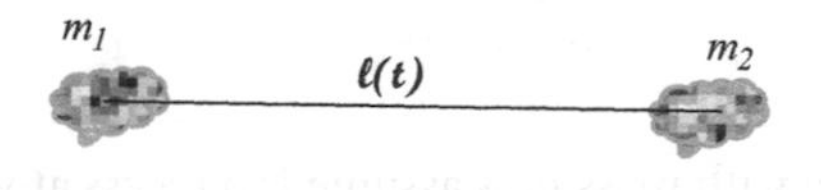

FIGURE 4.8 Space expands with time and the two distant galaxies appear to be recessing from each other.

$$\text{v} = \frac{d}{dt}l(t) = \dot{l}(t) = l_o\dot{a}(t) \tag{4.107}$$

To make the situation more general let us look at the motion of one galaxy of mass m outside a sphere containing a large number of galaxies such as in Figure 4.9.

We assume an isotropic distribution of galaxies within the sphere. The total mass is M with density ρ. In such a case, for the purpose of calculating the effect of gravity on the galaxy outside the sphere of radius l, we may consider all masses as concentrated at the center of the sphere with a total mass M equal to the sum of all galaxy masses within the sphere. We thus have for the galaxy of mass m outside the sphere:

$$Kinetic\ energy = \frac{1}{2}m\ \dot{l}^2 \tag{4.108}$$

$$Potential\ energy = -\frac{GMm}{l} \tag{4.109}$$

where G is the gravitational constant. The Lagrangian is thus:

$$L = \frac{1}{2}m\ \dot{l}^2 + \frac{GMm}{l} \tag{4.110}$$

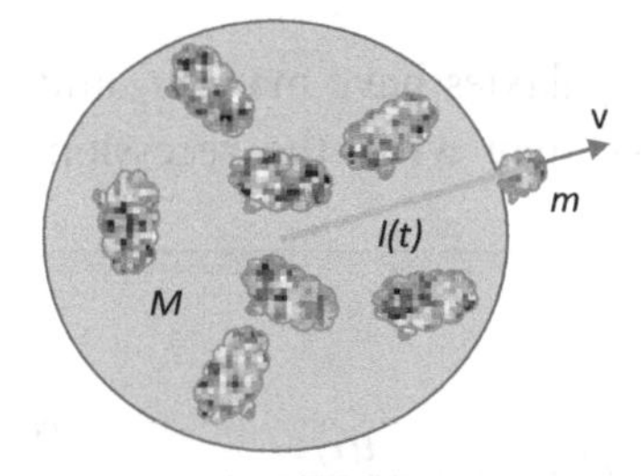

FIGURE 4.9 A galaxy with mass m is assumed to recess at velocity v from a large number of galaxies, forming a sphere whose total mass M is the sum of all galaxies within.

We use l, the radius of the sphere, as generalized coordinates, and Lagrange's equation for that dynamical variable is:

$$\frac{\partial L}{\partial l} - \frac{d}{dt}\frac{\partial L}{\partial \dot{l}} = 0 \tag{4.111}$$

We evaluate the two terms of Lagrange's equations thus:

$$\frac{\partial L}{\partial l} = -\frac{GMm}{l^2} \tag{4.112}$$

$$\frac{d}{dt}\frac{\partial L}{\partial \dot{l}} = \frac{d}{dt} m\dot{l} = m\ddot{l} \tag{4.113}$$

Consequently, we obtain:

$$\ddot{l} = -\frac{GM}{l^2} \tag{4.114}$$

However, the mass within the sphere is given by:

$$M = \frac{4}{3}\pi l^3 \rho \tag{4.115}$$

Thus, we have:

$$\frac{\ddot{l}}{l} = -\frac{4}{3}\pi G\rho \tag{4.116}$$

We use the relation $l(t) = a(t)l_o$ and obtain finally:

$$\frac{\ddot{a}}{a} = -\frac{4}{3}\pi G\rho \tag{4.117}$$

On the other hand, the total energy E is given by the Hamiltonian, H:

$$E = H = \dot{l}\frac{\partial L}{\partial \dot{l}} - L \tag{4.118}$$

We evaluate the various terms and obtain:

$$E = \frac{1}{2}m\dot{l}^2 - \frac{GMm}{l} \tag{4.119}$$

after some algebra using the definition of M in terms of density and

$$l(t) = a(t)l_o \tag{4.120}$$

$$\dot{l}(t) = \dot{a}(t)l_o \tag{4.121}$$

we finally obtain:

$$\left(\frac{\dot{a}}{a}\right)^2 = \frac{4}{3}G\pi\rho + \frac{2E}{ma^2l_o^2} \tag{4.122}$$

This equation gives the rate at which the galaxy recedes from the ensemble of galaxies within the sphere. This calculation can be made for any galaxy and a sphere with an arbitrary radius l. It can thus be interpreted as the rate at which the galaxies recede from each other. The term $\left(\frac{\dot{a}}{a}\right)$ is Hubble's expansion factor called $\mathcal{H}$. This factor is a function of time, the density continuously reducing with time and the last term itself being a function of the expansion parameter $a(t)$. Although we have not used Einstein field equations, we obtain two equations that describe the expansion of the universe as done in the context of general relativity. They give the same physical results as Friedmann's solution of Einstein field equations. The discussion can be extended to include the concept of critical density defined as:

$$\rho_c = \frac{3\mathcal{H}_o^2}{8\pi G} \tag{4.123}$$

We also define the parameter Ω:

$$\Omega = \frac{\rho}{\rho_c} \tag{4.124}$$

Eq. 4.123 becomes:

$$\Omega - 1 = -\frac{2E}{m\dot{l}_o^2} \tag{4.125}$$

As a speculation, we may interpret this result in terms of an equilibrium. If $\Omega = 1$ we have critical density and $E = 0$. The universe was born from nothing and will expand forever. The Big Bang concept appears as a creation of mass from nothing in a small global entity leading to negative potential gravitational energy. This energy is compensated by a positive kinetic energy, including rest mass, leading to fast expansion (inflation) at the *beginning* of its existence and resulting in the expansion rate that we know today. The term *beginning* raises questions since time, as we know it, did not exist before the so-called Big Bang.

H. ELECTROMAGNETIC FORCE

As is evident from the above applications, it is not always obvious what form the Lagrangian must take in order to represent a physical situation where a force is present. The validity of the Lagrangian chosen is often confirmed when the results obtained in the analysis using Lagrange's equations are in agreement with the known laws of nature. In the case of the motion of a particle of mass m at low velocity subjected to a force, the Lagrangian was introduced directly and defined as the difference between kinetic and potential energy. Actually, the definition was in a sense justified by means of d'Alembert's principle. It was then used in that form to calculate the dynamics of an object or particle by means of Lagrange's equation and found in agreement with Newton's principles. If the particle travels at high speed, however, it was not obvious what exact form the Lagrangian should take. We based our choice on the obligation that at low velocities the Lagrangian should be in agreement with the results obtained when Newton's principles are used with a mass independent of velocity. In fact, Newton's principles work so well at low velocity that a theory that would be in contradiction with them would immediately be rejected. A factor of the form $\sqrt{1-(\mathrm{v}/c)^2}$ was introduced in the kinetic energy term to take into account the effect of high velocity and to be compatible with the results obtained in the standard approach of the special theory of relativity. All those results essentially lead us to the understanding that the Lagrangian approach should give results that are compatible with classical results that we already know agree with experimental results.

In the case of a charged particle in an electric field, the force can be derived as the gradient of a scalar potential and the approach used is the same as the one used above for the case of a particle submitted to a conservative force. But what happens when the force is created by a magnetic field? The force is proportional to the velocity of the particle. In such a case, it is not obvious that our approach with a potential independent of velocity applies. However, it is possible to show that the force involved leads to a new form for the potential, leading to the possibility of continuing to use Lagrange's equations for analyzing the dynamics of the particle. We now proceed with this calculation.

We recall that electric and magnetic fields are described in space by Maxwell's equations:

$$\nabla \cdot \boldsymbol{E} = \frac{\rho}{\varepsilon_o} \tag{4.126}$$

$$\nabla \times \boldsymbol{E} = -\frac{\partial \boldsymbol{B}}{\partial t} \tag{4.127}$$

$$\nabla \cdot \boldsymbol{B} = \boldsymbol{0} \tag{4.128}$$

$$\nabla \times \boldsymbol{B} = \frac{\boldsymbol{j}}{\varepsilon_o c^2} + \frac{1}{c^2}\frac{\partial \boldsymbol{E}}{\partial t} \tag{4.129}$$

where $\boldsymbol{E}$ and $\boldsymbol{B}$ are the electric and magnetic (induction) fields present at position $\boldsymbol{r}$, ε_o is the vacuum electric constant and c is the speed of light. We recall that ρ is the electric charge density and $\boldsymbol{j}$ is the current density. Those equations are fundamental and their validity has been demonstrated to be in total agreement with experimental results. It was also established that the force on a particle with electric charge q, moving at velocity **v** in superposed electric and magnetic fields, is given by the Lorentz equation:

$$\boldsymbol{F} = q(\boldsymbol{E} + \mathbf{v} \times \boldsymbol{B}) \tag{4.130}$$

However, the field B may be derived from a vector potential A and is given by:

$$\boldsymbol{B} = \boldsymbol{\nabla} \times \boldsymbol{A} \tag{4.131}$$

We can replace this term in Eq. 4.127 and obtain:

$$\boldsymbol{\nabla} \times \left(\boldsymbol{E} + \frac{\partial \boldsymbol{A}}{\partial \boldsymbol{t}}\right) = 0 \tag{4.132}$$

As we have seen previously, in order for this equation to be valid, vector algebra implies that the term within parenthesis be the gradient of a potential ϕ that we write as $-\boldsymbol{\nabla}\phi$. We thus have:

$$E = -\boldsymbol{\nabla}\phi - \frac{\partial \boldsymbol{A}}{\partial t} \tag{4.133}$$

We replace this result into the force equation 4.130, making use also of Eq. 4.131, and we obtain:

$$\boldsymbol{F} = \boldsymbol{q}\left(-\boldsymbol{\nabla}\phi - \frac{\partial \boldsymbol{A}}{\partial \boldsymbol{t}} + \mathbf{v} \times \boldsymbol{\nabla} \times \boldsymbol{A}\right) \tag{4.134}$$

We now need to evaluate the term $\mathbf{v} \times \boldsymbol{\nabla} \times \boldsymbol{A}$ and $\frac{\partial A}{\partial t}$. This can be done using vector algebra as described by Goldstein (Classical Mechanics Section 1.5), by means of an evaluation of the form of the two terms for the Cartesian coordinate components. For example, for the x component, the force takes the form:

$$F_x = q\left\{-\frac{\partial}{\partial x}(\phi - \mathbf{v} \cdot \mathbf{A}) - \frac{d}{dt}\left(\frac{d}{dv_x}(\mathbf{v} \cdot \mathbf{A})\right)\right\} \tag{4.135}$$

We may then define a potential energy U as:

$$U = q\phi - q\boldsymbol{A} \cdot \mathbf{v} \tag{4.136}$$

making $\boldsymbol{F_x}$:

$$F_x = -\frac{\partial U}{\partial x} + \frac{d}{dt}\frac{\partial U}{\partial \mathrm{v}_x} \tag{4.137}$$

We encountered in the derivation of Lagrange's equation by means of d'Alembert's principle a similar situation where we had defined a generalized force. We can use the same approach and write the Lagrangian of interaction for particle α, having kinetic energy $\frac{1}{2}m_\alpha \dot{r}_\alpha^2$:

$$L = T - U = \frac{1}{2}m_\alpha \dot{r}_\alpha^2 - q_\alpha \phi(r_\alpha) + q_\alpha\, \boldsymbol{A}(\boldsymbol{r}_\alpha) \cdot \dot{\boldsymbol{r}}_\alpha \tag{4.138}$$

identifying the fields at position r_α of the particle of charge q_α and mass m_α.

We have thus obtained an expression for the Lagrangian describing the interaction of a particle moving at speed $\dot{r}_\alpha$ in a superposition of electric and magnetic fields at position r_α. This was made possible by means of a clear identification of the kinetic and potential energy of the particle in electric and magnetic fields. Since the interaction takes place at a point in space, because the particle is at position r_α, we can apply directly Lagrange's equation to that Lagrangian with dynamical variables r_α and $\dot{r}_\alpha$. We only need to evaluate the derivatives relative to r_α and $\dot{r}_\alpha$ and we should obtain a description of the motion of a particle in superposed electric and magnetic fields. This is readily done and, using vector algebra to evaluate $\boldsymbol{A}(\boldsymbol{r}_\alpha) \cdot \dot{\boldsymbol{r}}_\alpha$, a straightforward calculation gives the Lorentz force equation (see Cohen-Tannoudji et al. 1987, p. 105) confirming the validity of the least action principle and Lagrange's equations even in the case that the force is velocity dependent. In that last reference, the Lagrangian obtained above for a particle in motion in superposed electric and magnetic fields is assumed to have the form given by Eq. 4.138 and is the particle part of the so-called Standard Lagrangian. As will be discussed below, the total Lagrangian also includes fields $\boldsymbol{B}$ and $\boldsymbol{E}$ themselves. The validity of the Lagrangian is usually confirmed posteriorly through the demonstration summarized above by means of Lagrange's equations.

We have thus successfully applied the Lagrange approach to a particle of charge q and mass m in superposed electric and magnetic fields. We have determined the form of the Lagrangian that results in such a situation and using that Lagrangian in Lagrange's equation we came back to the Lorentz force expression, Eq. 4.130. Now we may ask ourselves questions about the fields themselves. We know that $\boldsymbol{E}$ and $\boldsymbol{B}$ contain energy and change with time. Can we write a Lagrangian for such fields? This is what we examine in the next chapter.

CHAPTER V

Fields and Quantum Physics

In our analysis, we have limited ourselves, up to now, to the mechanics of a particle or object with mass m in motion in space that has been submitted to known forces such as gravitational and electromagnetic forces. The particle studied was located at a position determined by means of an appropriate coordinate system. The number of degrees of freedom of the particle was determined by fixing the number of required coordinates. The pendulum, for example, had two constraints and only one degree of freedom. In that case the pendulum ball's location was determined by means of the angle θ. A particle in space without constraints has three degrees of freedom, say x, y and z, with t acting as a parameter. Its position is represented by $\boldsymbol{r}_\alpha$ $(x,\ y,\ z,\ t)$. Its speed is then given by $\dot{r}_\alpha(x,\ y,\ z, t)$. The Lagrangian can then be written as a simple expression. Actually, we could derive it using d'Alembert's principle as $L = T-V$. That Lagrangian could then be used in Lagrange's equation applied to each of the coordinates or degrees of freedom. The results obtained agreed with those obtained by means of Newton's principles, showing the validity of Hamilton's principle of least action when it was used in the definition of the action (Eq. 3.42). In those calculations, we were thus in the presence of a limited number of degrees of freedom or generalized coordinates.

However, as mentioned above, we may ask ourselves what happens when we deal with a field that is itself spread over a region of space? Should we be able to write a Lagrangian for such a field? As we just recalled, we arrived at the concept of the Lagrangian by introducing the principle of least action or d'Alembert's principle. When we deal with a particle or a mechanical system the word "action" seems to mean something familiar: we were dealing with an object moving in space under the influence of a force and there was an action taking place. We obtained results compatible with those obtained by means of Newton's principles and electromagnetic forces. The Lagrangian assumed was that which was able to fill that basic requirement. But in the case of fields, based on what requirement are we going to choose the form of the Lagrangian? It is not obvious at all. Fields are generally described by so-called field equations. The best example is probably Maxwell's equation for electric and magnetic fields. If they depend on time they are coupled to each other and lead to the generation of a wave described by a so-called wave equation. We must assume that the Lagrangian chosen for the fields involved must be an expression that leads to those fields and equations.

A. ELECTROMAGNETIC FIELD

As an example, let us look at the electric and magnetic (induction) fields $\boldsymbol{E}$ and $\boldsymbol{B}$. Above, we placed an electric charge moving in such fields. The charge interacted with those fields and was localized at position r_α with three degrees of freedom. We could identify, using some straightforward algebra, a potential energy term describing the interaction between the fields and the charge at position $\boldsymbol{r}_\alpha$. We could write the Lagrangian. But if we look at the field itself we see we have an entity that spreads over space. We can specify its value or amplitude at a given position, but that value changes continuously in terms of its position or coordinates and the field, varying in time, may take any amplitude at a given coordinate. We are essentially in the presence of a continuous function with an infinite number of degrees of freedom and we need to consider this when we analyze such fields by themselves. For example, we cannot simply talk about the energy in the field $\boldsymbol{E}$ and $\boldsymbol{B}$ at one point. We need to talk about their energy density at that point. This is given by:

$$\text{Electric energy density} = \frac{1}{2}\varepsilon_o E^2 \qquad (5.1)$$

$$\text{Magnetic energy density} = \frac{1}{2}\frac{B^2}{\mu_o} \tag{5.2}$$

The total energy over a region of space is then obtained by integration. This is very different from the case of a particle whose position could be written explicitly and its velocity could be determined at a point. In that case, kinetic and potential energy could be written explicitly at that position. The Lagrangian could then be written in terms of kinetic and potential energy, functions of coordinates and velocity and the amplitude of the fields at that point. But in the case studied here there are no particles.

Furthermore, what does action mean in that case? We know that those fields must obey Maxwell's equations. This is a good start in our analysis. It fixes some conditions on the relative behavior of electric and magnetic fields. We know how the fields vary in space and consequently, the action through the Lagrangian should be a function of space. If the fields are dependent on time we have a wave propagating in space and we know that the behavior of that wave derived from Maxwell's equations is in total agreement with observations. Consequently, the essential condition is that the Lagrangian assumed for the field, when studied with Lagrange's equation, gives Maxwell's equations as solution and leads to the same wave equation.

Thus, the whole discussion can be summarized by saying that, with fields, we are dealing with a continuous system rather than a discrete system like the one we encountered in the case of a particle in interaction with a field at a point in space. In such a case it is necessary to talk about density of the physical entity we deal with, as we did above with regard to energy. Since the Lagrangian is expressed in terms of energy, we may very well define it as:

$$L = \int \mathcal{L} dx\, dy\, dz \tag{5.3}$$

where script $\mathcal{L}$ is a density, which we will call the Lagrangian density. $\mathcal{L}$ is then a function of coordinates. It is of course function of the amplitude of the fields at the positions x, y, z, or other coordinates depending on the system used, as well as its derivative in respect to those coordinates since it varies over space. Electromagnetic fields, static or varying, are formed of electric and magnetic components, which can be represented, as recalled above, by potentials ϕ and $\boldsymbol{A}$, as in Eq. 4.133. We use the symbol $\mathcal{F}_j$ to represent those potentials, which can then be used as generalized coordinates for representing the fields. We thus write very generally the Lagrangian density as:

$$\mathcal{L} = \mathcal{L}\left(\mathcal{F}_j, \frac{\partial \mathcal{F}_j}{\partial x}, \frac{\partial \mathcal{F}_j}{\partial y}, \frac{\partial \mathcal{F}_j}{\partial z}, \frac{\partial \mathcal{F}_j}{\partial t}, t\right) \tag{5.4}$$

We have introduced as a variable the derivative of $\mathcal{F}_j$ in respect to coordinates similar to speed in the discrete case. We have also made possible a direct time dependence of $\mathcal{L}$.

As in the case of the particle dynamics, the action is then written in terms of those definitions as:

$$J = \int_1^2 \iiint \mathcal{L} d^3 r \, dt \tag{5.5}$$

where the limits of integration represent times t_1 and t_2, fixing the period over which the field's evolution is considered. The term $d^3 r$ represents an element of the volume over which the Lagrangian density $\mathcal{L}$ is integrated, as made explicit by means of the triple integral sign. On the other hand, the least action principle says that this action is extremal in the sense that it is stationary. In that case, a small variation of generalized coordinates in which $\mathcal{L}$ is expressed should not cause a change in the action J in first order. Thus, we may write, as we did earlier in the case of a particle action:

$$\delta J = \delta \int_1^2 \iiint d^3 r \, \mathcal{L} \, dt = 0 \tag{5.6}$$

We now need to express this condition explicitly. We note first that in the case of the discrete particle the definition of the action J was done in terms of a Lagrangian, which is itself a function of coordinates. We were in the presence of J, a function of functions L. J was called a functional. We are in a similar situation and here J is itself a functional.

Without going into semantics, we can do the same kind of analysis as we did earlier in the discrete case and attempt to solve Eq. 5.6 in a similar way. We need to express the variation of J as a variation of L. Since $\mathcal{L}$ is function of $\mathcal{F}_j$, we can write the change $\delta\mathcal{L}$ as:

$$\delta\mathcal{L} = \sum_{j,i}\left[\frac{\partial\mathcal{L}}{\partial\mathcal{F}_j}\partial\mathcal{F}_j + \frac{\partial\mathcal{L}}{\partial\dot{\mathcal{F}}_j}\partial\dot{\mathcal{F}}_j + \frac{\partial\mathcal{L}}{\partial\left(\frac{\partial\mathcal{F}_j}{\partial x_i}\right)}\partial\left(\frac{\partial\mathcal{F}_j}{\partial x_i}\right)\right] \tag{5.7}$$

This expression represents the variation of $\mathcal{L}$ in terms of the variation of the potentials used as generalized coordinates. We assume that the Lagrangian density is not dependent explicitly on time. We can replace this relation directly in Eq. 5.6. We thus have as final expression:

$$\delta J = \int_1^2\iiint\sum_{j,i}\left[\frac{\partial\mathcal{L}}{\partial\mathcal{F}_j}\partial\mathcal{F}_j + \frac{\partial\mathcal{L}}{\partial\dot{\mathcal{F}}_j}\partial\dot{\mathcal{F}}_j + \frac{\partial\mathcal{L}}{\partial\left(\frac{\partial\mathcal{F}_j}{\partial x_i}\right)}\partial\left(\frac{\partial\mathcal{F}_j}{\partial x_i}\right)\right]dt\,dx_1dx_2dx_3 = 0 \tag{5.8}$$

where d^3r has been written explicitly as $dx\,dy\,dz = dx_1dx_2dx_3$. Using a similar approach as that used in the discrete case, we can integrate by part the integral over time and after some straightforward algebra, similar to that used in that case, we obtain the condition that must be fulfilled by the Lagrangian density as (see Cohen-Tannoudji et al., 1987, p. 94; Goldstein, 1959, Section 11.2):

$$\sum_i \partial_i\frac{\partial\mathcal{L}}{\partial\left(\frac{\partial\mathcal{F}_j}{\partial x_i}\right)} + \frac{d}{dt}\frac{\partial\mathcal{L}}{\partial\dot{\mathcal{F}}_j} - \frac{\partial\mathcal{L}}{\partial\mathcal{F}_j} = 0 \qquad i = 1,\ 2,\ 3 \tag{5.9}$$

for each generalized coordinate $\mathcal{F}_j$. If the field is a complex quantity, a similar expression is derived for the complex part with $\mathcal{F}_j$ replaced by its complex conjugate $\mathcal{F}_j^*$. These are now Lagrange's equations which apply to the Lagrangian $\mathcal{L}$ of the electromagnetic field. This calculation does not tell us what form the Lagrangian density should have. It only tells us what conditions $\mathcal{L}$ must fulfill to be valid, that is, for example, give the same result as Maxwell's equations.

In the case of the electromagnetic field, the Lagrangian density that is proposed is a function of the field density and is written as:

$$\mathcal{L}_{field} = \frac{\mathcal{E}_o}{2}\left[E^2(r) - cB^2(r)\right] \tag{5.10}$$

The fields $\boldsymbol{B}$ and $\boldsymbol{E}$ are related by equations 4.131 and 4.133, which define $\boldsymbol{A}$ and ϕ respectively, and the scalar and vector potentials are used as generalized coordinates. In terms of this proposal, the total Lagrangian, including both the fields and a particle α in such fields, is then written in three parts:

$$L = L_{part.} + L_{field} + L_{int.} \tag{5.11}$$

where the various parts are identified as:

Particle Lagrangian:

$$L_{part} = \frac{1}{2} m_\alpha \dot{r}_\alpha^2 \tag{5.12}$$

Interaction Lagrangian:

$$L_{int} = -q_\alpha \phi(r_\alpha) + q_\alpha\, \boldsymbol{A}(r_\alpha) \cdot \dot{\boldsymbol{r}}_\alpha \tag{5.13}$$

Field Lagrangian:

$$L_{field} = \iiint d^3 r \frac{\mathcal{E}_o}{2}\left[E^2(r) - c^2 B^2(r)\right] \tag{5.14}$$

It is noted that we have limited the analysis to the interaction of a particle of mass m_α and electrical charge q_α. We can generalize the analysis to a distribution of charge of density ρ and a current density $\boldsymbol{j}$ and make it compatible with Maxwell's equations as written in Eqs. 4.126–4.129. In such a case the total Lagrangian becomes:

$$L = \frac{1}{2}\sum_{\alpha} m_{\alpha}\dot{r}_{\alpha}^{2} + \iiint(-\rho(r)\phi(r) + \boldsymbol{j}(r)\cdot\boldsymbol{A}(r))\,d^3\boldsymbol{r} + \iiint\frac{\varepsilon_o}{2}\left[E^2(r) - c^2B^2(r)\right]d^3r \tag{5.15}$$

It should be noted that the last term, L_{field}, in the total Lagrangian is rather different from the other two terms. It concerns the fields by themselves. The first term represents the kinetic energy of the particles while the second term represents the interaction between the particles and the fields. We can thus rewrite our Lagrangian density as:

$$\mathcal{L} = (-\rho(r)\phi(r) + \boldsymbol{j}(r)\cdot\boldsymbol{A}(r))\,d^3r + \frac{\epsilon_o}{2}\left[E^2(r) - c^2B^2(r)\right]d^3r \tag{5.16}$$

The fields exist over all space and take continuous values. We can use Eq. 4.131 and 4.133 to write $\boldsymbol{E}$ and $\boldsymbol{B}$ in terms of ϕ and $\boldsymbol{A}$. Those are definitions that can be used in our derivations. Is this Lagrangian $\mathcal{L}_{field}$ a valid representation of the field, obeying the least action principle? We have as before a way of checking its validity. We can use Lagrange's Eq. 5.9 to verify that the relation between $\boldsymbol{E}$, $\boldsymbol{B}$, ρ and $\boldsymbol{j}$ obey Maxwell's equations that we know have been demonstrated to be valid. Let us evaluate the various terms of Eq. 5.9. We use ϕ and $\boldsymbol{A}$ as dynamical variables and situate ourselves in a Cartesian space x, y, z. The kinetic energy of the particles does not enter in the calculation since in the Lagrangian it is not function of either ϕ or $\boldsymbol{A}$. We thus have in the case of ϕ and $\dot{\phi}$:

$$\frac{\partial\mathcal{L}}{\partial\dot{\phi}} = 0 \tag{5.17}$$

$$\frac{\partial\mathcal{L}}{\partial\phi} = -\rho \tag{5.18}$$

$$\partial_i\frac{\partial\mathcal{L}}{\partial(\partial_i\phi)} = \partial_i\frac{\partial\mathcal{L}}{\partial E_i}\frac{\partial E_i}{\partial(\partial_i\phi)} = \partial_i(-\varepsilon_o E_i) \tag{5.19}$$

We have similar expressions for all coordinates $i = x, y, z$ and the result is a summation over the three resulting Lagrange's equations that include three derivatives ∂_i. We finally obtain:

$$\nabla \cdot E = \frac{\rho}{\varepsilon_o} \tag{5.20}$$

which is Maxwell's first equation (Eq. 4.126). We can do the same exercise with $\boldsymbol{A}$ as a dynamical variable. The algebra is similar, and one obtains Maxwell's fourth equation written above as Eq 4.129.

The Lagrangian chosen, Eq. 5.16, can thus be considered appropriate. On that basis, we may very well conclude that, in principle, it should be possible to imagine a Lagrangian for an arbitrary field. An exercise such as the one done above can then be used to verify its validity and concordance with already known results, either theoretical or experimental.

B. GRAVITATIONAL FIELD

The gravitational field $\boldsymbol{E}_{gr}$, which we introduced earlier, can be used as an example. Actually, we have obtained an expression for the field as (Eq. 2.94):

$$\boldsymbol{\nabla} \cdot \boldsymbol{E}_{gr} = -4\pi G\rho \tag{5.21}$$

where ρ is the density of matter and the field is defined as the gradient of a potential ϕ, which gives rise to Poisson's equation:

$$\boldsymbol{\nabla}^2 \phi = 4\pi G\rho \tag{5.22}$$

We thus have an equation very similar to that of Maxwell for the electric field. With some imagination we can write the Lagrangian for the gravitational field as:

$$\mathcal{L} = -\rho\phi - \frac{1}{8\pi G} E_{gr}^2 = -\rho\phi - \frac{1}{8\pi G} (\boldsymbol{\nabla}\phi)^2 \tag{5.23}$$

We can use that Lagrangian in Lagrange's equation and do the same analysis as the one we have done for the electromagnetic field. We obtain Eq. 5.22, which is in complete agreement with Newton's principles.

C. A WORD ABOUT MATTER WAVES AND QUANTUM MECHANICS

Although the intention of this text is to provide the reader with an introduction to the use of the least action principle and Lagrange's approach in classical physics, it is worth mentioning its use in other fields also, such as in wave mechanics. In fact, it is probably in that field that the Lagrangian approach has been most successful in describing how our universe functions in the atomic and subatomic world and in making a rational connection between the quantum and the classical worlds.

In wave mechanics, particle properties are described by a so-called wave representing the classical matter field ψ used in Schrödinger's approach of wave mechanics. The Lagrangian density of that field is assumed to have the form (Cohen-Tannoudji et al., 1987, p. 159):

$$\mathcal{L} = \frac{i\hbar}{2}(\psi^* \dot{\psi} - \dot{\psi}^* \psi) - \frac{\hbar^2}{2m} \nabla \psi^* \cdot \nabla \psi - V(r) \psi^* \psi \tag{5.24}$$

where ψ is a complex field. It is the wave function introduced by Schrödinger in wave mechanics. In the present context of Lagrange's approach, the functions ψ and ψ^* are used as generalized coordinates. The various terms of Lagrange's equations 5.9 are then evaluated using that somewhat complex Lagrangian. Such an exercise leads to Schrödinger's equation:

$$i\hbar \frac{\partial \psi}{\partial t} = \left[V(r) - \frac{\hbar^2}{2m} \nabla^2 \right] \psi \tag{5.25}$$

We note that the functions ψ and ψ^* are classical fields. The system can then be quantized by replacing those functions by operators ψ and ψ^* and the use of quantum mechanical commutator rules. The exercise leads to the concept of second quantization and the introduction of bosons and fermions fields.

On the other hand, it is possible to provide a new formulation of quantum mechanics by assuming that a phase is associated with the path followed by a particle in its evolution from a point 1 to a point 2 (see Feynman, "Nobel Lecture" in *La nature de la Physique* 1980). It is assumed that all paths between those points are possible. Those paths are characterized by a probability amplitude called $K_n(2,1)$, which is just a measure of the probability that the particle starting at point 1 reaches point 2 by means of a path identified as n. We then associate a phase to that amplitude and postulate that the probability amplitude of path n is given by:

$$K_n(2,1) = Ce^{i\frac{J_n}{\hbar}} \tag{5.26}$$

where C is a normalization constant and J_n is the classical action characterizing path n and defined earlier by means of Eqs. 3.43, 5.3 and 5.5 as:

$$J_n = \int_1^2 L(r, \dot{r}, t)dt \tag{5.27}$$

It is then postulated that the actual path followed by the particle is determined by the sum over all path probabilities K_n. Since the phase given by the exponent $\frac{iJ_n}{\hbar}$ varies from one path to the other, there will be interferences and terms will cancel in the summation. As pointed out by Feynman (1980, Nobel Lecture), the above postulates lead directly to Schrödinger's equation and justify in a sense the use of the classical Lagrangian in the multiple paths approach used. On the other hand, it is worth mentioning a most important aspect of that approach. It is observed that the phase is given by the ratio $\frac{J_n}{\hbar}$. If J_n is very large compared to $\hbar$, the phase varies rapidly from one path to the other, except when the paths are close to the actual classical path. In that case, the action is extremal and is of the order of $\hbar$. This observation identifies in a sense the limit between classical and quantum mechanics and justifies again the validity of the least action principle.

Finally, it should be mentioned that Lagrangians have also been established for many other types of interactions such as the weak and strong forces (See for example Lahiri and Pal 2005; Wilczek 2000). Those Lagrangians have a very complex form.

There are many other fields of physics and many mechanical systems to which the Lagrangian approach can be applied. The goal of the present text is to introduce the beginner to the least action principle and the Lagrangian approach in an elementary way. We have thus limited the presentation to a limited number of applications. Any reader who is interested in more advanced applications may find some in the list of volumes given at the end of this monography and on the Internet.

CHAPTER VI

Conclusion

One may ask again, what is the point of using the Lagrangian approach rather than directly using classical equations? When we use the proper Lagrangian in classical mechanics we obtain, from Lagrange's equation, Newton's principles and the appropriate trajectory in a gravitational field. If we apply Lagrange's equation to the appropriate Lagrangian for a charge in an electromagnetic field, we obtain Lorentz force. If we used the Lagrangian proposed in Chapter V for electromagnetic fields, we obtain Maxwell's equations. We can introduce relativity by defining an appropriate Lagrangian. There are many other examples that we could introduce in other subfields of physics where the analysis could be made through the use of the Lagrangian approach. The exercise appears to be in the finding of the appropriate Lagrangian, which, when used with Lagrange's equation, leads to the laws that we already know. In general, it is observed that the form of the Lagrangian that is required for the results to be compatible with already known laws is not obvious at all. The Lagrangian correctness is essentially demonstrated *a posteriori*.

From the examples chosen, however, it appears first that the solution of various problems using the Lagrange approach is simpler because, when we know the Lagrangian, we deal only with scalars and not with vectors. Secondly, it appears that we could deduct a new principle from a deeper analysis of the properties of Lagrange's equation: symmetry introduces conservation laws. This is a very important result and, as pointed out by Feynman, the law that associates a conservation property to symmetry is probably one of the most profound things in physics (Feynman et al. 1965).

On the other hand, the whole theory developed here for mechanical systems can be applied to fields in general in the atomic world and standard quantization rules can be used. A Lagrangian can be proposed and analyzed as we did above. The solution of Lagrange's equations with that Lagrangian leads then to fundamental equations such as Schrödinger's equation describing the quantum world. The approach, when coupled to special relativity, has led to the development of quantum field theory, which has been extremely successful in describing the evolution of systems at the quantum level. Furthermore, the approach has been used in the case where the physics connected to the presence of weak and strong forces is studied (see Wilczek, *Physics Today* 2000). Unification based on symmetry has resulted. Furthermore, new fields can be imagined and studied, such as the Higgs field giving mass to particles. The Lagrangian approach offers a simple approach to determine the characteristics of those fields and their validity in addressing a particular observation.

A most important aspect, however, concerns the validity of the principle of least action itself being used as the basis of the development of the Lagrangian and Hamilton approaches. In that context, we may speculate that this principle, (which claims that the action defined by means of the Lagrangian is stationary), is universal. It seems that nature in its evolution follows that principle within the context of our mathematics. It can be seen as unifying, in a sense, all forces and interactions that we observe in the universe. The least action principle thus appears to be one of the most basic rules regulating the evolution of the universe. Furthermore, it provides an avenue through which we better understand the limit between quantum mechanics and classical physics. This is a rather inspiring observation.

The least action principle does not tell us why the universe reacts the way it does. Nevertheless, it seems to tell us that the universe acts as to maintain equilibrium and to take a path that is optimized. We must recall that the whole theory is developed with our tools, mathematics and calculus, a language that the human being has invented and that, we must confess, is still characterized by the limits of our intelligence. Is nature using a similar language? We do not know. Is that principle as expressed in our mathematical language a valid and complete representation of reality? We do not know. Is it an approximation of a more complete and deeper principle? Most probably. There are obviously doubts about an approach, such as the one just outlined with regard to predetermined evolution and totally predictable characteristics. In the universe, there are multitudes of objects and they follow many possible paths; chaos seems to be present. In practice, we address the resulting complexity by means of statistics. On the other hand, quantum theory is based on the uncertainty principle and the trajectory of an electron or of a so-called elementary particle is not determined *a priori* (see, for example, Trinh Xuan Thuan et al. 2008). The interference of electrons travelling through a double slit is a most convincing observation of a trajectory *a priori* undetermined.

We are part of that universe; we do not have a god's view of the ensemble and we don't know where that universe comes from. Nevertheless, the discovery of a broad-spectrum principle, such as the least action principle, using the tool of reasoning that has been given to us by evolution, has provided us with a most interesting insight into the inner workings of the universe we live in. Even though we may never know what this whole thing is about, we should be very proud of that achievement.

Selected physics constants that are used in this text

Speed of light:	$c = 299\,792\,458$ (exactly) m/s
Elementary charge:	$e = 1.6022 \times 10^{-19}$ C
Electron mass:	$m_e = 9.109 \times 10^{-31}$ kg
Proton mass :	$m_p = 1.672 \times 10^{-27}$ kg
Planck's constant:	$h = 6.6261 \times 10^{-34}$ J s
Vacuum dielectric constant:	$\varepsilon_o = 1/\mu_o c^2 = 8.854\,187\,817 \times 10^{-12}$ F/m
Magnetic constant:	$\mu_o = 4\pi \times 10^{-7}$(exactly) N/A^2
Universal gravitational constant:	$G = 6.674 \times 10^{-11}$ m^3/kg s^2
Distance Sun–Earth:	$D_{se} = 149.6 \times 10^6$ km
Earth's mass:	$M_e = 5.972 \times 10^{24}$ kg
Earth's radius:	$Re = 6371$ km
Earth's gravitational constant:	$g_{earth} = 9.80$ m/s^2
Sun's mass:	$M_s = 1.989 \times 10^{30}$ kg
Distance Moon–Earth:	$D_{me} = 384\,400$ km
Moon's mass:	$M_m = 7.3476 \times 10^{22}$ kg
Moon's radius:	$R_m = 1737$ km
Moon's gravitational constant:	$g_m = 1.62$ m/s^2

References

A First Book of Quantum Field Theory, A. Lahiri, P. Pal, Alpha Science Int, 2005.

A Universe from Nothing, L.M. Krauss, Free Press, 2012.

Classical Mechanics, H.G. Goldstein, Addison Wesley, 1959.

La nature de la Physique, Richard Feynman, éditions du Seuil, 1980.

Le monde s'est-il créé tout seul?, Trinh Xuan Thuan, Ilya Prigogine, Albert Jacquard, Joël de Rosnay, Jean-Marie Pelt, Henri Atlan, Édition Alain Michel, 2008.

Mécanique quantique, C. Cohen-Tannoudji, B. Diu, F. Laloë, Hermann, 1973.

Photons et atomes, Claude Cohen-Tannoudji, Jacques Dupont-Roc, Gilbert Grynberg, Editions du CNRS, 1987.

Principles of Electricity and Magnetism, G.P. Harnwell, McGraw Hill Book Co, 1949.

Principles of Physical Cosmology, P.J.E. Peebles, Princeton University Press, 1993.

QCD Made Simple, F. Wilczek, Physics Today, p. 22, 2000.

The Feynman Lectures on Physics, R.P. Feynman, R.R. Leighton, M. Sands, Addison Wesley, 1965.

The Quantum Physics of Atomic Frequency Standards, J. Vanier, C. Audoin, Adam-Hilger, 1989.

The Universe, A Challenge to the Mind, Jacques Vanier, Imperial College Press, 2011.

Theoretical Physics, Georg Joos, Ira M. Freeman, Blackie & Son Lim, 1954.

Index